HEINEMANN MODULAR MATHEMATICS

for

EDEXCEL AS AND A-LEVEL

Further Pure Mathematics 3

Geoff Mannall   Michael Kenwood

Endorsed by **edexcel**

heinemann.co.uk

✓ Free online support
✓ Useful weblinks
✓ 24 hour online ordering

01865 888058

Heinemann

Inspiring generations

Heinemann Educational Publishers,
Halley Court, Jordan Hill, Oxford OX2 8EJ
Part of Harcourt Education

Heinemann is the registered trademark of Harcourt Education Limited

First published 2005

09 08 07 06 05
10 9 8 7 6 5 4 3 2

10-digit ISBN: 0 435511 02 5

13-digit ISBN: 978 0 435511 02 9

Cover design by Gecko Limited

Original design by Geoffrey Wadsley; additional design work by Jim Turner

Typeset and illustrated by Tech-Set Limited, Gateshead, Tyne and Wear

Printed and bound in the UK by Scotprint

Acknowledgements:

The publisher's and authors' thanks are due to Edexcel for permission to
reproduce questions from past examination papers. These are marked with an [E].
    The answers have been provided by the authors and are not the responsibility
of the examining board.

About this book

This book is designed to provide you with the best preparation possible for your Edexcel FP3 exam. The series authors are senior examiners and exam moderators themselves and have a good understanding of Edexcel's requirements.

Finding your way around

To help to find your way around when you are studying and revising use the:

- **edge marks** (shown on the front page) – these help you to get to the right chapter quickly;
- **contents list** – this lists the headings that identify key syllabus ideas covered in the book so you can turn straight to them;
- **index** – if you need to find a topic the **bold** number shows where to find the main entry on a topic.

Remembering key ideas

We have provided clear explanations of the key ideas and techniques you need throughout the book. Key ideas you need to remember are listed in a **summary of key points** at the end of each chapter and marked like this in the chapters:

■ $$e^{i\theta} = \cos\theta + i\sin\theta$$

Exercises and exam questions

In this book questions are carefully graded so they increase in difficulty and gradually bring you up to exam standard.

- **past exam questions** are marked with an [E];
- **review exercise** on page 165 helps you practise answering questions from several areas of mathematics at once, as in the real exam;
- **exam style practice paper** – this is designed to help you prepare for the exam itself;
- **answers** are included at the end of the book – use them to check your work.

Contents

Maclaurin and Taylor series

1

In Book C2, chapter 7, you met the infinite geometric series

$$a + ar + ar^2 + \ldots = a(1 - r)^{-1}, |r| < 1$$

In Book C2, chapter 5, you met the infinite binomial series

$$1 + nx + \frac{n(n-1)}{2!}x^2 + \frac{n(n-1)(n-2)}{3!}x^3 + \ldots = (1+x)^n$$

where n is rational and $|x| < 1$.

Such infinite series are called **power series**. In this chapter you will learn how to express other function such as e^x, $\sin x$ and $\ln(1+x)$ as power series.

1.1 Higher derivatives

In Book C2, chapter 9, the notation for second and third derivatives was given as $\dfrac{d^2y}{dx^2}$ and $\dfrac{d^3y}{dx^3}$. For a function f given by the equation $y = f(x)$, the 1st, 2nd, 3rd ... nth derivatives are either denoted by

$$\frac{dy}{dx}, \frac{d^2y}{dx^2}, \frac{d^3y}{dx^3}, \ldots, \frac{d^ny}{dx^n}$$

or by

$$f'(x), f''(x), f'''(x), \ldots, f^{(n)}(x)$$

Example 1

Given that $y = e^{x^2}$, find $\dfrac{dy}{dx}$, $\dfrac{d^2y}{dx^2}$, and $\dfrac{d^3y}{dx^3}$.

$\dfrac{dy}{dx} = \dfrac{d}{dx}(e^{x^2}) = e^{x^2}\dfrac{d}{dx}(x^2)$, using the chain rule of differentiation.

So:
$$\frac{dy}{dx} = 2xe^{x^2}$$

$$\frac{d^2y}{dx^2} = \frac{d}{dx}\left(\frac{dy}{dx}\right) = \frac{d}{dx}(2xe^{x^2})$$

Using the product rule of differentiation you get:

$$\frac{d^2 y}{dx^2} = e^{x^2} \frac{d}{dx}(2x) + 2x \frac{d}{dx}(e^{x^2})$$

$$= 2e^{x^2} + 2x(2xe^{x^2})$$

$$= 2e^{x^2} + 4x^2 e^{x^2}$$

$$= 2e^{x^2}(1 + 2x^2)$$

$$\frac{d^3 y}{dx^3} = \frac{d}{dx}\left(\frac{d^2 y}{dx^2}\right)$$

$$= \frac{d}{dx}[2e^{x^2}(1 + 2x^2)]$$

$$= (1 + 2x^2)\frac{d}{dx}(2e^{x^2}) + 2e^{x^2}\frac{d}{dx}(1 + 2x^2)$$

$$= 4x(1 + 2x^2)e^{x^2} + 2e^{x^2}(4x)$$

$$= (4x + 8x^3 + 8x)e^{x^2}$$

$$= 4xe^{x^2}(3 + 2x^2)$$

1.2 Maclaurin's series

Consider the function f given by $f(x) \equiv (1 + x)^n$, and suppose that $f(x)$ can also be expressed as

$$f(x) \equiv A_0 + A_1 x + A_2 x^2 + A_3 x^3 + \cdots + A_r x^r + \cdots$$

where $A_0, A_1, A_2, A_3, \cdots, A_r, \cdots$ are constants.

At $x = 0$: $\qquad\qquad (1 + x)^n = (1 + 0)^n = 1$

Also, at $x = 0$: $\quad A_0 + A_1 x + A_2 x^2 + \cdots = A_0$

So: $\qquad\qquad\qquad\qquad A_0 = 1$

Differentiating $f(x)$ with respect to x you obtain

$$f'(x) \equiv A_1 + 2A_2 x + 3A_3 x^2 + \cdots + rA_r x^{r-1} + \cdots$$

But $\qquad\qquad\qquad f'(x) = n(1 + x)^{n-1}$

So at $x = 0$, you have:

$$f'(x) = A_1 = n(1 + 0)^{n-1} \Rightarrow A_1 = n$$

Differentiating both expressions for $f'(x)$:

$$f''(x) \equiv 2A_2 + 3(2)A_3 x + \cdots + r(r - 1)A_r x^{r-2} + \cdots$$

Also: $\qquad\qquad f''(x) = n(n - 1)(1 + x)^{n-2}$

At $x = 0$, $2A_2 = n(n-1)(1+0)^{n-2} \Rightarrow A_2 = \dfrac{n(n-1)}{2}$

Continuing the differentiating and equating gives:

$$A_3 = \frac{n(n-1)(n-2)}{3!} \text{ and } A_r = \frac{n(n-1)(n-2)\cdots(n-r+1)}{r!}$$

Hence:

$$(1+x)^n = 1 + nx + \frac{n(n-1)}{2!}x^2 + \frac{n(n-1)(n-2)}{3!}x^3 + \cdots + \frac{n(n-1)(n-2)\cdots(n-r+1)}{r!}x^r + \cdots$$

which you know from Book C2 is a binomial series. In arriving at this result, you should appreciate that it has been *assumed* that it is valid mathematically to differentiate an infinite series term by term. You cannot prove this assumption at present.

■ **Generally, for the continuous function f, given by**

$$\mathbf{f : x \mapsto f(x), \ x \in \mathbb{R}}$$

if f(0), f′(0), f″(0), $\cdots$ f$^{(r)}$(0), $\cdots$ all have finite values, then

$$\mathbf{f(x) = f(0) + xf'(0) + \frac{x^2}{2!}f''(0) + \cdots + \frac{x^r}{r!}f^{(r)}(0) + \cdots}$$

This series is known as the **Maclaurin expansion of f(x) in ascending powers of x**.

For a given function the series may converge to f(x) for *all* values of x. However, frequently the expansion only holds for a restricted range of values of x.

Example 2

Expand e^x in ascending powers of x.

$$f(x) = e^x, \ f'(x) = e^x, \ f''(x) = e^x, \cdots, f^{(r)}(x) = e^x$$

At $x = 0$: $f(0) = f'(0) = f''(0) = f'''(0) = \cdots = f^{(r)}(0) = 1$

Hence, using Maclaurin's series you have:

$$e^x = 1 + x + \frac{x^2}{2!} + \frac{x^3}{3!} + \cdots + \frac{x^r}{r!} + \cdots$$

The expansion is valid for all values of x.

Example 3

Expand $\cos x$ in ascending powers of x.

Let $f(x) = \cos x$

$$f'(x) = -\sin x, \ f''(x) = -\cos x, \ f'''(x) = \sin x, \ f''''(x) = \cos x$$

and you then have: $f(0) = 1$, $f'(0) = 0$, $f''(0) = -1$, $f'''(0) = 0$, $f''''(0) = 1$ and the cycle of four values 1, 0, −1, 0 repeats itself. So you have from Maclaurin's expansion:

$$\cos x = 1 - \frac{x^2}{2!} + \frac{x^4}{4!} - \cdots + (-1)^r \frac{x^{2r}}{(2r)!} + \cdots$$

This expansion is valid for all values of x.

Example 4

Expand $\ln(1 + x)$ in ascending powers of x.

$$f(x) = \ln(1 + x)$$

$$f'(x) = \frac{1}{1 + x} = (1 + x)^{-1}; \quad f''(x) = -(1 + x)^{-2};$$

$$f'''(x) = (-1)(-2)(1 + x)^{-3}; \quad f''''(x) = (-1)(-2)(-3)(1 + x)^{-4}$$

and $\qquad f^{(r)}(x) = (-1)(-2)(-3)\cdots(-r + 1)(1 + x)^{-r}$

So $f(0) = 0$, $f'(0) = 1$, $f''(0) = -1$, $f'''(0) = 2!$, $f''''(0) = -3! \cdots$
$f^{(r)}(0) = (-1)^{r+1}(r - 1)!$

So the Maclaurin expansion is

$$\ln(1 + x) = x - \frac{x^2}{2!} + \frac{2!x^3}{3!} - \frac{3!x^4}{4!} + \cdots + (\frac{-1)^{r+1}(r - 1)!x^r}{r!} + \cdots$$

$$= x - \frac{x^2}{2} + \frac{x^3}{3} - \frac{x^4}{4} + \cdots + \frac{(-1)^{r+1}x^r}{r} + \cdots$$

This expansion is valid for $-1 < x \leqslant 1$.

The series for $\sin x$ is set as an exercise below. The series is valid for all values of x.

You should learn the conditions of validity for these series.

Exercise 1A

Using Maclaurin's expansion, and differentiation, show that:

1 $e^{-x} = 1 - x + \frac{x^2}{2!} - \frac{x^3}{3!} + \cdots + (-1)^r \frac{x^r}{r!} + \cdots$

2 $(1 - x)^{-1} = 1 + x + x^2 + x^3 + \cdots + x^r + \cdots$

3 $e^{2x} = 1 + 2x + 2x^2 + \frac{4}{3}x^3 + \cdots + \frac{2^r x^r}{r!} + \cdots$

4 $\sin x = x - \frac{x^3}{3!} + \frac{x^5}{5!} - \cdots + (-1)^r \frac{x^{2r+1}}{(2r + 1)!} + \cdots$

5 $\ln(1 - x) = -x - \frac{x^2}{2} - \frac{x^3}{3} - \cdots - \frac{x^r}{r} - \cdots$

Find the first three non-zero terms in the Maclaurin expansion of the function given in ascending powers of x:

6 $\tan x$ **7** $\sin^2 x$ **8** $\ln\left(\dfrac{1+x}{1-x}\right)$, $|x| < 1$

9 $(1 - 2x^2)^{\frac{1}{2}}$ **10** $e^x \cos x$ **11** $\sinh x$

12 $\sin x \cosh x$ **13** $\cosh^2 x$ **14** $\dfrac{x}{1+x^2}$, $|x| < 1$ **15** $e^{x^2 - x}$

1.3 Using the polynomial series form of functions to find approximations for the functions

When x is small, it is evident that x^2, $x^3 \cdots$ get successively smaller. For example:

$$x = \tfrac{1}{10}, \quad x^2 = \tfrac{1}{100}, \quad x^6 = \tfrac{1}{1\,000\,000}, \text{ etc.}$$

By using this fact, you can take an approximation of the polynomial form of a function to represent the function, if x is sufficiently small. For example, you know that

$$\sin x = x - \frac{x^3}{3!} + \frac{x^5}{5!} - \cdots$$

$$\cos x = 1 - \frac{x^2}{2!} + \frac{x^4}{4!} - \cdots$$

So if you take terms in x^3 and higher powers of x to be negligible, then

∎ $\sin x \approx x$ and $\cos x \approx 1 - \dfrac{x^2}{2}$

Also in Exercise 1A, question 6, you should have found that:

$$\tan x = x + \tfrac{1}{3}x^3 + \tfrac{2}{15}x^5 \cdots$$

So you can also say, for small x, that

∎ $\tan x \approx x$

These three approximations have widespread applications in science, technology and engineering, together with approximations from the binomial series such as

∎ $(1 + x)^{\frac{1}{2}} \approx 1 + \tfrac{1}{2}x - \tfrac{1}{8}x^2$

Example 5
Find a quadratic polynomial approximation for $\dfrac{\sin 2x}{1 + x}$, given that x is small.

$$\frac{\sin 2x}{1+x} = \sin 2x(1+x)^{-1}$$

You can say that: $\qquad \sin 2x \approx 2x$

$$(1+x)^{-1} = 1 + (-1)(x) + \frac{(-1)(-2)}{2!}(x)^2 + \cdots$$

$$= 1 - x + x^2 - \cdots$$

So: $\qquad \sin 2x(1+x)^{-1} \approx 2x(1 - x + x^2)$

$$\approx 2x - 2x^2$$

(x^3 and higher powers of x being negligible).

Example 6

Given that x is small, show that $\dfrac{3\sin x}{2 + \cos x} \approx x$.

$$3\sin x = 3\left(x - \frac{x^3}{3!} + \cdots\right)$$

$$(2 + \cos x)^{-1} \approx \left(2 + 1 - \frac{x^2}{2}\right)^{-1}$$

$$= 3^{-1}\left(1 - \frac{x^2}{6}\right)^{-1}$$

$$= 3^{-1}\left(1 + \frac{x^2}{6} + \cdots\right) \text{ (binomial series)}$$

So: $\qquad \dfrac{3\sin x}{2 + \cos x} \approx 3\left(x - \frac{x^3}{6}\right)(3^{-1})\left(1 + \frac{x^2}{6}\right)$

$$= x + \frac{x^3}{6} - \frac{x^3}{6} - \frac{x^5}{36}$$

$$= x\left(1 - \frac{x^4}{36}\right)$$

$$\approx x, \text{ as required.}$$

Example 7

Show that $\lim\limits_{x \to 0}\left(\dfrac{1 - \cos 4x + x\sin 3x}{x^2}\right) = 11$.

Using the approximations for small x:

$$\cos 4x \approx 1 - \frac{(4x)^2}{2} \quad \text{and} \quad \sin 3x \approx 3x$$

you have:

$$\frac{1 - \cos 4x + x \sin 3x}{x^2} \approx \frac{1 - (1 - 8x^2) + 3x^2}{x^2}$$

$$= 11$$

Notice that other terms would have x^2, x^4, etc. in them, so that:

$$\lim_{x \to 0} \left(\frac{1 - \cos 4x + x \sin 3x}{x^2} \right) = 11, \text{ as required.}$$

Exercise 1B

1 Given that x is small, find the constants A and B such that

$$(x + \sin x) \cos x \approx Ax + Bx^3$$

2 Given that x is small, find the constants C and D such that

$$\tan x \approx Cx + Dx^3$$

3 Find $\displaystyle\lim_{x \to 0} \left(\frac{\sin\left(\frac{\pi}{6} + x\right) - \sin\frac{\pi}{6}}{\sin 2x} \right)$.

4 Given that x is small, show that

$$\frac{\sin x - x \cos x}{x^3} \approx \tfrac{1}{3}$$

5 Given that x is so small that terms in x^3 and higher powers of x may be disregarded, show that

$$\ln\left[\frac{(1 + 2x)^2}{1 - 3x}\right] = 7x + \tfrac{1}{2}x^2$$

6 Show that for small x:

$$\frac{(1 + x)^{\frac{1}{2}}}{(1 - x)^2} \approx 1 + \tfrac{5}{2}x + \tfrac{31}{8}x^2$$

7 Given that x takes a value near $\frac{\pi}{2}$, explain why $\cos x \approx \frac{\pi}{2} - x$. Use this approximation to find (to 2 decimal places) the smallest positive root of the equation

$$\cos x = \frac{x}{10}$$

8 Given that x is small, show that

$$e^{\sin x} = 1 + x + \tfrac{1}{2}x^2 + Ax^3$$

and determine the value of A. You may assume that terms in x^4 and higher powers of x can be disregarded.

9 Evaluate: (a) $\lim\limits_{x \to 0} \left(\dfrac{\sin x - x}{\sin x - x \cos x} \right)$ (b) $\lim\limits_{x \to 0} \left(\dfrac{\ln(1 + x) - x}{\sin^2 x} \right)$.

10 Given that x is small and that terms in x^4 and higher powers of x may be disregarded, show that

$$\ln(\sec x + \tan x) = x + \tfrac{1}{6}x^3$$

1.4 Taylor's series

Maclaurin's expansion for f(x) is

$$f(x) = f(0) + xf'(0) + \frac{x^2}{2!}f''(0) + \frac{x^3}{3!}f'''(0) + \cdots + \frac{x^r}{r!}f^{(r)}(0) + \cdots$$

It is used to expand functions such as $\sin x$, $\ln(1 + x)$, e^x, $\tanh x$ and $\operatorname{arsinh} x$, to mention just a few. Series expansions have many uses and applications, one of which is approximating a function by the first few terms of its series expansion. You have seen this already in section 1.3 of this chapter. Maclaurin's series can be adapted to a more general form if you want to approximate to the value of f(x) when $x \approx a$, where $a \neq 0$. These adapted forms of Maclaurin's expansion are called **Taylor expansions** or **Taylor series**.

Consider the functions f and g, where $f(x + a) \equiv g(x)$, $a \neq 0$ and a is a constant.

By Maclaurin's series:

$$g(x) = g(0) + xg'(0) + \frac{x^2}{2!}g''(0) + \frac{x^3}{3!}g'''(0) + \cdots + \frac{x^r}{r!}g^{(r)}(0) + \cdots$$

So: $f(x + a) = f(a) + xf'(a) + \dfrac{x^2}{2!}f''(a) + \dfrac{x^3}{3!}f'''(a) + \cdots + \dfrac{x^r}{r!}f^{(r)}(a) + \cdots$

because if $f(x + a) = g(x)$ then $f^{(r)}(x + a) = g^{(r)}(x)$, for $r = 1, 2, 3, \ldots$ and this implies that $f^{(r)}(a) = g^{(r)}(0)$ for $r = 1, 2, 3, \ldots$

■ **The expansion**

$$\mathbf{f(x + a) = f(a) + xf'(a) + \frac{x^2}{2!}f''(a) + \cdots + \frac{x^r}{r!}f^{(r)}(a) + \cdots}$$

is known as Taylor's series (or Taylor's expansion).

In this series, replacing x by $x - a$ gives you the alternative form of Taylor's series.

■ $f(x) = f(a) + (x-a)f'(a) + \dfrac{(x-a)^2}{2!}f''(a) + \cdots + \dfrac{(x-a)^r}{r!}f^{(r)}(a) + \cdots$

where $f(x)$ is given as a series in ascending powers of $(x-a)$. The convergence of Maclaurin and Taylor series is beyond your syllabus but you may assume that any series you are asked to find or use is valid for the work you are asked to do with it.

Example 8
Expand $\ln x$ in ascending powers of $(x-1)$ up to and including the term in $(x-1)^3$.

Differentiate with respect to x:

Let $f(x) = \ln x$, then $f'(x) = \dfrac{1}{x}$, $f''(x) = -\dfrac{1}{x^2}$, $f'''(x) = \dfrac{2}{x^3}$

Substitute $x = 1$:

$f(1) = \ln 1 = 0$, $f'(1) = 1$, $f''(1) = -1$, $f'''(1) = 2$

Using Taylor's expansion in the form

$f(x) = f(a) + (x-a)f'(a) + \dfrac{(x-a)^2}{2!}f''(a) + \dfrac{(x-a)^3}{3!}f'''(a) + \cdots$

you obtain when $a = 1$ and $f(x) = \ln x$:

$\ln x = \ln 1 + (x-1)(1) + \dfrac{(x-1)^2}{2!}(-1) + \dfrac{(x-1)^3}{3!}(2) + \cdots$

That is:

$\ln x = (x-1) - \tfrac{1}{2}(x-1)^2 + \tfrac{1}{3}(x-1)^3 + \cdots$

1.5 The Taylor series method for series solutions of differential equations

Suppose you are asked to solve the differential equation

$$\frac{dy}{dx} = f(x, y)$$

where $f(x, y)$ is an expression in x and y, and you know further that $y = y_0$ at $x = x_0$. In the majority of cases which you will meet, $x_0 = 0$ but this is not necessarily so (see Example 11, page 12).

By direct substitution in the differential equation of $y = y_0$ and $x = x_0$ you can find $f(x_0, y_0)$, which is the value of $\dfrac{dy}{dx}$ at (x_0, y_0).

We use the notation $\left(\dfrac{dy}{dx}\right)_{x_0}$, $\left(\dfrac{d^2y}{dx^2}\right)_{x_0}$, $\left(\dfrac{d^3y}{dx^3}\right)_{x_0}$, $\ldots$ for the values of $\dfrac{dy}{dx}, \dfrac{d^2y}{dx^2}, \dfrac{d^3y}{dx^3}, \ldots$ at $x = x_0$. Once you have established the

values of $\left(\dfrac{dy}{dx}\right)_{x_0}, \left(\dfrac{d^2y}{dx^2}\right)_{x_0}, \left(\dfrac{d^3y}{dx^3}\right)_{x_0}, \ldots$ by further differentiation

of, and substitution in, the differential equation, the series solution of the differential equation can be found by using Taylor's series in the form

$$y = y_0 + (x - x_0)\left(\frac{dy}{dx}\right)_{x_0} + \frac{(x - x_0)^2}{2!}\left(\frac{d^2y}{dx^2}\right)_{x_0} + \frac{(x - x_0)^3}{3!}\left(\frac{d^3y}{dx^3}\right)_{x_0} + \cdots$$

Notice that, if $x_0 = 0$, the series solution is simply

$$y = y_0 + x\left(\frac{dy}{dx}\right)_0 + \frac{x^2}{2!}\left(\frac{d^2y}{dx^2}\right)_0 + \frac{x^3}{3!}\left(\frac{d^3y}{dx^3}\right)_0 + \cdots$$

which is, of course, Maclaurin's series. The following examples are typical and should clarify the methods for you.

Example 9

Use the Taylor series method to find the series solution of the differential equation $\dfrac{dy}{dx} = x^2 + y^2$, for which $y = 1$ at $x = 0$, up to and including the term in x^3.

Starting with $\dfrac{dy}{dx} = x^2 + y^2$, you have by repeated differentiation with respect to x:

$$\frac{d^2y}{dx^2} = 2x + 2y\frac{dy}{dx} = 2x + 2y(x^2 + y^2)$$

$$= 2x + 2x^2y + 2y^3$$

$$\frac{d^3y}{dx^3} = 2x + 4xy + 2x^2\frac{dy}{dx} + 6y^2\frac{dy}{dx}$$

$$= 2 + 4xy + 2x^2(x^2 + y^2) + 6y^2(x^2 + y^2)$$

$$= 2 + 4xy + 2x^4 + 8x^2y^2 + 6y^4$$

Hence at $x = x_0 = 0$, $y_0 = 1$ you have:

$$\left(\frac{dy}{dx}\right)_0 = 0^2 + 1^2 = 1$$

$$\left(\frac{d^2y}{dx^2}\right)_0 = 2(0) + 2(0)(1) + 2(1)^3 = 2$$

$$\left(\frac{d^3y}{dx^3}\right)_0 = 2 + 4(0)(1) + 2(0) + 8(0)^2(1)^2 + 6(1)^4 = 8$$

Using Taylor's series in the form where $x_0 = 0$, you have

$$y = y_0 + x\left(\frac{dy}{dx}\right)_0 + \frac{x^2}{2!}\left(\frac{d^2y}{dx^2}\right)_0 + \frac{x^3}{3!}\left(\frac{d^3y}{dx^3}\right)_0 + \cdots$$

and $y = 1 + x + x^2 + \frac{8}{6}x^3$ is the series solution of the differential equation up to and including the term in x^3.

So the solution required is $y = 1 + x + x^2 + \frac{4}{3}x^3$.

You can make the method of solution of the differential equation shorter by using

$$\frac{d^2y}{dx^2} = 2x + 2y\frac{dy}{dx}$$

$$\frac{d^3y}{dx^3} = 2 + 2\left(\frac{dy}{dx}\right)^2 + 2y\frac{d^2y}{dx^2}$$

Now you know that $y_0 = 1$, $\left(\frac{dy}{dx}\right)_0 = 1$ and using these at $x_0 = 0$:

$$\left(\frac{d^2y}{dx^2}\right)_0 = 2x_0 + 2y_0\left(\frac{dy}{dx}\right)_0 = 2$$

$$\left(\frac{d^3y}{dx^3}\right)_0 = 2 + 2\left(\frac{dy}{dx}\right)_0^2 + 2y_0\left(\frac{d^2y}{dx^2}\right)_0$$
$$= 2 + 2 + 4$$
$$= 8$$

Then proceed, as before, to form the series solution.

Example 10

Use the Taylor series method to find the series solution, up to the term in x^4, of the differential equation

$$\frac{d^2y}{dx^2} + 2y\frac{dy}{dx} + y^2 = x$$

given that $\frac{dy}{dx} = y = 1$ at $x = 0$.

Using the standard notation, you have at $x_0 = 0$, $y_0 = 1$ and $\left(\frac{dy}{dx}\right)_0 = 1$.

From the differential equation at $x = x_0 = 0$:

$$\left(\frac{d^2y}{dx^2}\right)_0 + 2y_0\left(\frac{dy}{dx}\right)_0 + y_0^2 = x_0$$

That is: $\left(\frac{d^2y}{dx^2}\right)_0 + 2(1)(1) + 1^2 = 0 \Rightarrow \left(\frac{d^2y}{dx^2}\right)_0 = -3$

Differentiating the differential equation with respect to x:

$$\frac{d^3y}{dx^3} + 2\left(\frac{dy}{dx}\right)^2 + 2y\frac{d^2y}{dx^2} + 2y\frac{dy}{dx} = 1$$

At $x = x_0$: $\left(\frac{d^3y}{dx^3}\right)_0 + 2\left(\frac{dy}{dx}\right)_0^2 + 2y_0\left(\frac{d^2y}{dx^2}\right)_0 + 2y_0\left(\frac{dy}{dx}\right)_0 = 1$

That is: $\left(\frac{d^3y}{dx^3}\right)_0 + 2(1)^2 + 2(1)(-3) + 2(1)(1) = 1$

$$\Rightarrow \quad \left(\frac{d^3y}{dx^3}\right)_0 = 3$$

Differentiating again, with respect to x:

$$\frac{d^4y}{dx^4} + 4\frac{dy}{dx}\frac{d^2y}{dx^2} + 2\frac{dy}{dx}\frac{d^2y}{dx^2} + 2y\frac{d^3y}{dx^3} + 2\left(\frac{dy}{dx}\right)^2 + 2y\frac{d^2y}{dx^2} = 0$$

At $x = x_0$, using the values of y_0, $\left(\frac{dy}{dx}\right)_0$, $\left(\frac{d^2y}{dx^2}\right)_0$ and $\left(\frac{d^3y}{dx^3}\right)_0$ you have

$$\left(\frac{d^4y}{dx^4}\right)_0 + 4(1)(-3) + 2(1)(-3) + 2(1)(3) + 2(1)^2 + 2(1)(-3) = 0$$

$$\Rightarrow \quad \left(\frac{d^4y}{dx^4}\right)_0 = 16$$

Using the Taylor series in the form where $x_0 = 0$ you have

$$y = y_0 + x\left(\frac{dy}{dx}\right)_0 + \frac{x^2}{2!}\left(\frac{d^2y}{dx^2}\right)_0 + \frac{x^3}{3!}\left(\frac{d^3y}{dx^3}\right)_0 + \frac{x^4}{4!}\left(\frac{d^4y}{dx^4}\right)_0 \cdots$$

and so

$$y = 1 + x - \tfrac{3}{2}x^2 + \tfrac{1}{2}x^3 + \tfrac{2}{3}x^4 + \cdots$$

is the series solution of the differential equation, as required.

Example 11

$$\frac{d^2y}{dx^2} = xy + \frac{dy}{dx}$$

Given that $\frac{dy}{dx} = 2$ and $y = 1$ at $x = 1$, find the values of $\frac{d^2y}{dx^2}$ and $\frac{d^3y}{dx^3}$ at $x = 1$. Hence find the solution of the differential equation as a series in ascending powers of $(x - 1)$, up to and including the term in $(x - 1)^3$.

Differentiate $\dfrac{\mathrm{d}^2 y}{\mathrm{d}x^2} = xy + \dfrac{\mathrm{d}y}{\mathrm{d}x}$ with respect to x and you have

$$\frac{\mathrm{d}^3 y}{\mathrm{d}x^3} = y + x\frac{\mathrm{d}y}{\mathrm{d}x} + \frac{\mathrm{d}^2 y}{\mathrm{d}x^2}$$

Since $\dfrac{\mathrm{d}y}{\mathrm{d}x} = 2$ and $y = 1$ at $x = 1$, you have $\left(\dfrac{\mathrm{d}y}{\mathrm{d}x}\right)_1 = 2$ and $y_1 = 1$.

At $x = 1$, $\qquad \left(\dfrac{\mathrm{d}^2 y}{\mathrm{d}x^2}\right)_1 = y_1 + \left(\dfrac{\mathrm{d}y}{\mathrm{d}x}\right)_1 = 1 + 2 = 3$

and $\qquad \left(\dfrac{\mathrm{d}^3 y}{\mathrm{d}x^3}\right)_1 = y_1 + \left(\dfrac{\mathrm{d}y}{\mathrm{d}x}\right)_1 + \left(\dfrac{\mathrm{d}^2 y}{\mathrm{d}x^2}\right)_1 = 1 + 2 + 3 = 6$

Using Taylor's series in the form

$$y = y_1 + (x-1)\left(\frac{\mathrm{d}y}{\mathrm{d}x}\right)_1 + \tfrac{1}{2}(x-1)^2\left(\frac{\mathrm{d}^2 y}{\mathrm{d}x^2}\right)_1 + \tfrac{1}{6}(x-1)^3\left(\frac{\mathrm{d}^3 y}{\mathrm{d}x^3}\right)_1 + \cdots$$

where $y_1 = 1$, $\left(\dfrac{\mathrm{d}y}{\mathrm{d}x}\right)_1 = 2$, $\left(\dfrac{\mathrm{d}^2 y}{\mathrm{d}x^2}\right)_1 = 3$, and $\left(\dfrac{\mathrm{d}^3 y}{\mathrm{d}x^3}\right)_1 = 6$,

you obtain

$$y = 1 + 2(x-1) + \tfrac{3}{2}(x-1)^2 + (x-1)^3 + \cdots$$

as the series solution in ascending powers of $x - 1$ of the differential equation, as required.

1.6 Using series expansions in integration

You can sometimes find an approximate value for a definite integral by expanding the integrand $\mathrm{f}(x)$ in an infinite series and then integrating the series term by term. The success of this method depends on the limits of the integral being such that the series converges rapidly when you consider just a few terms, while the remaining terms are so small that they are negligible. Here is an example.

Example 12

Estimate the value of $\displaystyle\int_0^{0.6} \mathrm{e}^{-x^2}\,\mathrm{d}x$, giving your answer to 3 decimal places.

The series for e^x is:

$$\mathrm{e}^x = 1 + x + \frac{1}{2!}x^2 + \frac{1}{3!}x^3 + \cdots$$

If you replace x by $-x^2$ you get:

$$e^{-x^2} = 1 - x^2 + \frac{1}{2!}x^4 - \frac{1}{3!}x^6 + \cdots$$

Now consider just the first four terms and assume that integration of the series for e^{-x^2} term by term is valid. So:

$$\int_0^{0.6} e^{-x^2}dx = \int_0^{0.6} (1 - x^2 + \tfrac{1}{2}x^4 - \tfrac{1}{6}x^6 + \cdots)\,dx$$

$$= \left[x - \tfrac{1}{3}x^3 + \tfrac{1}{10}x^5 - \tfrac{1}{42}x^7 + \cdots \right]_0^{0.6}$$

$$= 0.6 - 0.072 + 0.007\,776 - 0.000\,667 \ldots$$

$$= 0.5351 \text{ (4 decimal places)}$$

That is:
$$\int_0^{0.6} e^{-x^2}dx = 0.535 \text{ (3 decimal places)}$$

You can see that, in this case, successive terms are rapidly getting smaller and smaller.

In cases like this a series expansion followed by integration of the first few terms of the series can give an acceptable estimate for such a definite integral. The trapezium rule could provide an alternative strategy for obtaining an estimate of the integral.

Exercise 1C

1 Given that x is small, show that
   (a) $e^{\sin x} \approx 1 + x + \frac{1}{2}x^2$
   (b) $\arcsin x \approx x + \frac{1}{6}x^3$
   when terms in x^4 and higher powers of x are neglected.

2 Expand $x^{\frac{1}{2}}$ using Taylor's series in ascending powers of $(x - 1)$ up to and including the term in $(x - 1)^3$.

3 Using Taylor's series expand (a) e^x (b) x^{-1}, in ascending powers of $(x - h)$ up to and including the term in $(x - h)^2$.

4 Using Taylor's series expand $x^2 \ln x$ in ascending powers of $(x - 1)$ up to and including the term in $(x - 1)^3$.

5 Using Taylor's series, expand $\sin(x + \frac{\pi}{6})$ in ascending powers of x up to and including the term in x^3.

6 Using Taylor's series, expand $\cos(x + \frac{\pi}{3})$ in ascending powers of x up to and including the term in x^3.

In questions 7–15, find the series solution in ascending powers of x, up to and including the term in x^3, of the differential equation:

7 $\dfrac{dy}{dx} = x^2 - y^2$ for which $y = 1$ at $x = 0$

8 $\dfrac{dy}{dx} = \frac{1}{2}y^2 - x^2$ for which $y = 2$ at $x = 0$

9 $y\dfrac{dy}{dx} = x \cosh y$ for which $y = 1$ at $x = 0$

10 $\dfrac{dy}{dx} = y \cos x$ for which $y = 1$ at $x = 0$

11 $\dfrac{d^2y}{dx^2} = xy + 1$ for which $\dfrac{dy}{dx} = 0$ and $y = 1$ at $x = 0$

12 $\dfrac{d^2y}{dx^2} + x\dfrac{dy}{dx} + y = 0$ for which $\dfrac{dy}{dx} = 0$ and $y = 1$ at $x = 0$

13 $\cos x \dfrac{dy}{dx} + 2y = 0$ for which $y = 1$ at $x = 0$

14 $\dfrac{d^2y}{dx^2} + 2\tan y \left(\dfrac{dy}{dx}\right)^2 = 0$ for which $y = \frac{\pi}{4}$ and $\dfrac{dy}{dx} = \frac{1}{2}$ at $x = 0$

15 $\dfrac{d^2y}{dx^2} + y\dfrac{dy}{dx} + y^3 = 0$ for which $\dfrac{dy}{dx} = y = 1$ at $x = 0$

16   Given that $\dfrac{d^2y}{dx^2} = x^2 + y^2$ and that $y = \dfrac{dy}{dx} = 1$ at $x = 1$, find a series solution for y in ascending powers of $(x - 1)$ up to and including the term in $(x - 1)^3$.

17   The differential equation $\dfrac{d^2y}{dx^2} + y\dfrac{dy}{dx} = x$ has $y = 0$ and $\dfrac{dy}{dx} = 1$ at $x = 1$.
Find a series solution in ascending powers of $(x - 1)$ up to and including the terms in $(x - 1)^3$.

18 $\dfrac{dy}{dx} = e^{xy}$ and $y = 1$ at $x = 0$.
Find the series solution of this differential equation in ascending powers of x up to and including the term in x^3.

19   Use the first three non-zero terms in the series expansion of $\sinh x^2$ in ascending powers of x to estimate $\displaystyle\int_0^{0.3} \sinh x^2 \, dx$, giving your answer to 7 decimal places.

20 Use the first four non-zero terms in the series expansion of $\cos x$ to estimate the value of $\displaystyle\int_0^{0.2} \cos x \, dx$.

21 Use the binomial series to find the first three non-zero terms in the expansion of $(1 - x^2)^{\frac{1}{3}}$ in ascending powers of x.

Giving your answer to 4 decimal places, estimate the value of
$$\int_0^{0.4} (1 - x^2)^{\frac{1}{3}} \, dx$$
by using your series.

SUMMARY OF KEY POINTS

1 If $y = f(x)$, successive differentiation with respect to x gives:

$$\frac{dy}{dx} = f'(x), \quad \frac{d^2y}{dx^2} = f''(x), \cdots, \frac{d^ny}{dx^n} = f^{(n)}(x)$$

2 Maclaurin's expansion:

$$f(x) = f(0) + \frac{x}{1!}f'(0) + \frac{x^2}{2!}f''(0) + \cdots + \frac{x^r}{r!}f^{(r)}(0) + \cdots$$

3 $\sin x = x - \dfrac{x^3}{3!} + \dfrac{x^5}{5!} - \cdots + (-1)^r \dfrac{x^{2r+1}}{(2r+1)!} + \cdots$

4 $\cos x = 1 - \dfrac{x^2}{2!} + \dfrac{x^4}{4!} - \cdots + (-1)^r \dfrac{x^{2r}}{(2r)!} + \cdots$

5 $\ln(1+x) = x - \dfrac{x^2}{2} + \dfrac{x^3}{3} - \cdots + (-1)^{r+1}\dfrac{x^r}{r} + \cdots$ $(-1 < x \leqslant 1)$

6 $e^x = 1 + x + \dfrac{x^2}{2!} + \dfrac{x^3}{3!} + \cdots + \dfrac{x^r}{r!} + \cdots$

For sufficiently small x, where x^3 and higher powers of x are disregarded:

7 $\sin x \approx x \approx \tan x$

8 $\cos x \approx 1 - \frac{1}{2}x^2$

9 $(1+x)^{\frac{1}{2}} \approx 1 + \frac{1}{2}x - \frac{1}{8}x^2$

10 $\ln(1+x) \approx x - \frac{1}{2}x^2$

11 $e^x \approx 1 + x + \frac{1}{2}x^2$

12 Taylor's series for $f(x + a)$ in ascending powers of x is

$$f(x + a) = f(a) + xf'(a) + \frac{x^2}{2!}f''(a) + \cdots + \frac{x^r}{r!}f^{(r)}(a) + \cdots$$

where a is constant.

13 Taylor's series for $f(x)$ in ascending powers of $(x - a)$ is

$$f(x) = f(a) + (x - a)f'(a) + \frac{(x - a)^2}{2!}f''(a) +$$

$$\cdots + \frac{(x - a)^r}{r!}f^{(r)}(a) + \cdots$$

where a is a constant.

14 The Taylor series solution of the differential equation $\frac{dy}{dx} = f(x, y)$ for which $y = y_0$ at $x = x_0$ is

$$y = y_0 + (x - x_0)\left(\frac{dy}{dx}\right)_{x_0} + \frac{(x - x_0)^2}{2!}\left(\frac{d^2y}{dx^2}\right)_{x_0}$$

$$+ \frac{(x - x_0)^3}{3!}\left(\frac{d^3y}{dx^3}\right)_{x_0} + \cdots$$

where $\left(\dfrac{d^ny}{dx^n}\right)_{x_0}$ is the value of $\dfrac{d^ny}{dx^n}$ at $x = x_0$.

Often $x_0 = 0$, and the solution is

$$y = y_0 + x\left(\frac{dy}{dx}\right)_0 + \frac{x^2}{2!}\left(\frac{d^2y}{dx^2}\right)_0 + \frac{x^3}{3!}\left(\frac{d^3y}{dx^3}\right)_0 + \cdots$$

15 An approximate value of a definite integral can sometimes be found by expanding the integrand in a series and then integrating the series term by term. This only works satisfactorily when the limits are substituted, if successive terms after the first few are rapidly getting smaller and smaller.

Complex numbers

2

The idea of an imaginary number and hence a complex number was introduced in Book FP1. Chapter 3 of that book describes how a complex number can be written in the form $a + ib$, where $a, b \in \mathbb{R}$, and shows how you can find the modulus and the argument of a complex number.

This chapter starts by showing that you can write a complex number in exponential form. You will then be introduced to de Moivre's theorem and hence how to find roots of a number, and then the relationship between trigonometric and hyperbolic functions. You will also learn about loci in the complex plane and inequalities involving complex numbers. The final section deals with some elementary transformations that involve complex numbers.

2.1 Exponential form of a complex number

In chapter 1 we showed that the series for the exponential function, e^x, $x \in \mathbb{R}$, is given by

$$e^x = 1 + x + \frac{x^2}{2!} + \frac{x^3}{3!} + \frac{x^4}{4!} + \cdots + \frac{x^r}{r!} + \cdots$$

Now it can be proved, although you will not find the proof here, that this series also holds if x is a complex number.

If you expand $e^{i\theta}$ as a series you obtain:

$$e^{i\theta} = 1 + i\theta + \frac{(i\theta)^2}{2!} + \frac{(i\theta)^3}{3!} + \frac{(i\theta)^4}{4!} + \cdots + \frac{(i\theta)^r}{r!} + \cdots$$

$$= 1 + i\theta - \frac{\theta^2}{2!} - \frac{i\theta^3}{3!} + \frac{\theta^4}{4!} + \cdots + \frac{i^r\theta^r}{r!} + \cdots$$

$$= 1 - \frac{\theta^2}{2!} + \frac{\theta^4}{4!} - \frac{\theta^6}{6!} + \frac{\theta^8}{8!} + \cdots + \frac{(-1)^r\theta^{2r}}{(2r)!} + \cdots$$

$$+ i\left(\theta - \frac{\theta^3}{3!} + \frac{\theta^5}{5!} - \frac{\theta^7}{7!} + \cdots + \frac{(-1)^r\theta^{2r+1}}{(2r+1)!} + \cdots\right)$$

If you remember the series for $\cos\theta$ and $\sin\theta$ (page 5) you will see that:

■ $$e^{i\theta} = \cos\theta + i\sin\theta$$

This is a useful result that you should remember.

Now the modulus–argument form of a complex number is

$$z = r(\cos\theta + i\sin\theta)$$

So you can now write a complex number in a third form, namely

■ $$z = r\,e^{i\theta}$$

where $r = |z|$ and $\theta = \arg z$.

This is called the exponential form of a complex number z.

2.2 Relations between trigonometric functions and hyperbolic functions

Now that you know that the exponential series is true for any complex number z, and that $e^{i\theta} = \cos\theta + i\sin\theta$, you can write

$$e^{iz} = \cos z + i\sin z$$

and, if you replace z by $-z$,

$$e^{i(-z)} = \cos(-z) + i\sin(-z)$$

Now
$$\cos(-z) = 1 - \frac{(-z)^2}{2!} + \frac{(-z)^4}{4!} - \frac{(-z)^6}{6!} + \cdots$$

$$= 1 - \frac{z^2}{2!} + \frac{z^4}{4!} - \frac{z^6}{6!} + \cdots = \cos z$$

and
$$\sin(-z) = (-z) - \frac{(-z)^3}{3!} + \frac{(-z)^5}{5!} - \frac{(-z)^7}{7!} + \cdots$$

$$= -z + \frac{z^3}{3!} - \frac{z^5}{5!} + \frac{z^7}{7!} - \cdots = -\sin z$$

So:
$$e^{-iz} = \cos z - i\sin z$$

and
$$e^{iz} + e^{-iz} = \cos z + i\sin z + (\cos z - i\sin z)$$
$$= 2\cos z$$

That is:

■ $$\cos z = \frac{e^{iz} + e^{-iz}}{2}$$

Also:
$$e^{iz} - e^{-iz} = \cos z + i\sin z - (\cos z - i\sin z)$$
$$= 2i\sin z$$

So:

■
$$\sin z = \frac{e^{iz} - e^{-iz}}{2i}$$

In Book FP2, chapter 1, the hyperbolic functions cosh and sinh were defined for real values and you may now assume that:

$$\cosh z = \frac{e^z + e^{-z}}{2}, \ z \in \mathbb{C}$$

and
$$\sinh z = \frac{e^z - e^{-z}}{2}, \ z \in \mathbb{C}$$

If you replace z by iz in the formula for $\cos z$ above, you get:

$$\cos iz = \frac{e^{i(iz)} + e^{-i(iz)}}{2}$$

$$= \frac{e^{i^2 z} + e^{-i^2 z}}{2}$$

$$= \frac{e^{-z} + e^z}{2}$$

$$= \cosh z$$

Similarly:

$$\sin iz = \frac{e^{i(iz)} - e^{-i(iz)}}{2i}$$

$$= \frac{e^{-z} - e^z}{2i}$$

$$= \frac{e^{-z} - e^z}{2i} \times \frac{i}{i}$$

$$= i\left(\frac{e^{-z} - e^z}{-2}\right)$$

$$= i\left(\frac{e^z - e^{-z}}{2}\right)$$

$$= i \sinh z$$

Also:
$$\cosh iz = \frac{e^{iz} + e^{-iz}}{2}$$

$$= \frac{(\cos z + i \sin z) + (\cos z - i \sin z)}{2}$$

$$= \frac{2\cos z}{2}$$

$$= \cos z$$

and $\qquad \sinh iz = \dfrac{e^{iz} - e^{-iz}}{2}$

$$= \frac{(\cos z + i \sin z) - (\cos z - i \sin z)}{2}$$

$$= \frac{2i \sin z}{2}$$

$$= i \sin z$$

These, then, are the relationships between the trigonometric and hyperbolic functions which you must remember:

■

$$\cos iz = \cosh z$$
$$\sin iz = i \sinh z$$
$$\cosh iz = \cos z$$
$$\sinh iz = i \sin z$$

The relations $\qquad \cos \theta = \cosh z$

$$\sin \theta = i \sinh z$$

where $iz = \theta$, show you a justification for Osborn's rule, which provides a means of obtaining identities for hyperbolic functions from identities for trigonometric functions (see Book FP2, page 6).

In an identity you can see that a term containing $\cos^2 \theta$ can be replaced by $\cosh^2 z$ and a term containing $\sin^2 \theta$ can be replaced by $(i \sinh z)^2 = -\sinh^2 z$. So we have, for example,

$$\cos^2 \theta + \sin^2 \theta \equiv 1 \quad \Rightarrow \quad \cosh^2 z + (i \sinh z)^2 \equiv 1$$

$$\Rightarrow \quad \cosh^2 z - \sinh^2 z \equiv 1$$

and a similar approach can be used for other identities.

Example 1

Express $z = 5 + 12i$ in the form (a) $r(\cos \theta + i \sin \theta)$ (b) $re^{i\theta}$ where $r = |z|$ and $\theta = \arg z$.

$$z = 5 + 12i$$

$$r = |z| = \sqrt{(5^2 + 12^2)} = \sqrt{(25 + 144)}$$

so: $\qquad\qquad r = 13$

$$\tan \theta = \tfrac{12}{5} \Rightarrow \theta = 1.18^c \ (3 \text{ s.f.})$$

So (a): $\qquad\qquad z = 13(\cos 1.18 + i \sin 1.18)$

or (b): $\qquad\qquad z = 13e^{1.18i}$

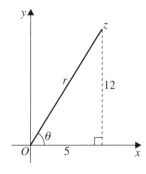

Example 2

Express $e^{-\frac{i\pi}{4}}$ in the form
  (a) $\cos \theta + i \sin \theta$, $-\pi < \theta \leqslant \pi$
  (b) $a + ib$, $a, b \in \mathbb{R}$

(a) $e^{-\frac{i\pi}{4}} = \cos\left(-\frac{\pi}{4}\right) + i\sin\left(-\frac{\pi}{4}\right) = \cos\frac{\pi}{4} - i\sin\frac{\pi}{4}$

    since $\cos(-\theta) = \cos\theta$ and $\sin(-\theta) = -\sin\theta$.

(b) Also, since $e^{-\frac{i\pi}{4}} = \cos\frac{\pi}{4} - i\sin\frac{\pi}{4}$

then:
$$e^{-\frac{i\pi}{4}} = \frac{1}{\sqrt{2}} - i\frac{1}{\sqrt{2}}$$

$$= \frac{1}{\sqrt{2}}(1 - i)$$

$$= \frac{\sqrt{2}(1 - i)}{2}$$

Example 3

Given that
$$\cos(z + w) \equiv \cos z \cos w - \sin z \sin w, \qquad z, w \in \mathbb{C}$$

prove that
$$\cosh(z + w) \equiv \cosh z \cosh w + \sinh z \sinh w$$

Since $\qquad\qquad\qquad \cos iz = \cosh z$

you have $\qquad\qquad \cos i(z + w) = \cosh(z + w)$

Thus: $\qquad \cosh(z + w) = \cos i(z + w)$

$$= \cos(iz + iw)$$

$$= \cos(iz)\cos(iw) - \sin(iz)\sin(iw)$$

$$= \cosh z \cosh w - (i\sinh z)(i\sinh w)$$

$$= \cosh z \cosh w - i^2 \sinh z \sinh w$$

So: $\qquad \cosh(z + w) \equiv \cosh z \cosh w + \sinh z \sinh w$

2.3 Multiplying and dividing two complex numbers

If $\qquad\qquad\qquad z_1 = r_1(\cos\theta_1 + i\sin\theta_1)$

and $\qquad\qquad\qquad z_2 = r_2(\cos\theta_2 + i\sin\theta_2)$

then: $\quad z_1 z_2 = (r_1\cos\theta_1 + ir_1\sin\theta_1)(r_2\cos\theta_2 + ir_2\sin\theta_2)$

$$= r_1 r_2 \cos\theta_1 \cos\theta_2 + ir_1 r_2 \cos\theta_1 \sin\theta_2 + ir_1 r_2 \sin\theta_1 \cos\theta_2 - r_1 r_2 \sin\theta_1 \sin\theta_2$$

$$= r_1 r_2 (\cos\theta_1 \cos\theta_2 - \sin\theta_1 \sin\theta_2) + ir_1 r_2 (\sin\theta_1 \cos\theta_2 + \cos\theta_1 \sin\theta_2)$$

$$= r_1 r_2 \cos(\theta_1 + \theta_2) + ir_1 r_2 \sin(\theta_1 + \theta_2)$$

$$= r_1 r_2 [\cos(\theta_1 + \theta_2) + i\sin(\theta_1 + \theta_2)]$$

You can picture what happens geometrically if you draw an Argand diagram:

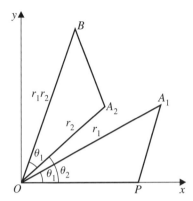

A_1 represents the number z_1.
A_2 represents the number z_2.
B represents the product z_1z_2.
You can see that the length of OB is $r_1 \times r_2$.

You can also see that

since $\angle A_1OP = \theta_1 = \arg z_1$

$\qquad \angle A_2OP = \theta_2 = \arg z_2$

and $\quad \angle BOP = \theta_1 + \theta_2 = \arg(z_1z_2)$

then $\qquad\qquad\qquad\qquad \angle BOA_2 = \theta_1$

If, in addition, the point P has coordinates $(1, 0)$, that is P represents the real number 1, then since $\angle BOA_2 = \angle A_1OP (= \theta_1)$ and since

$$\frac{OB}{OA_2} = \frac{r_1r_2}{r_2} = \frac{r_1}{1}$$

and

$$\frac{OA_1}{OP} = \frac{r_1}{1} = \frac{OB}{OA_2}$$

the triangles OBA_2 and OA_1P are similar.

So when you multiply two complex numbers, you *multiply* their moduli and you *add* their arguments.

Similarly, if

$$z_1 = r_1(\cos \theta_1 + i \sin \theta_1)$$

and

$$z_2 = r_2(\cos \theta_2 + i \sin \theta_2)$$

then

$$\frac{z_1}{z_2} = \frac{r_1 \cos \theta_1 + ir_1 \sin \theta_1}{r_2 \cos \theta_2 + ir_2 \sin \theta_2}$$

$$= \frac{(r_1 \cos \theta_1 + ir_1 \sin \theta_1)(r_2 \cos \theta_2 - ir_2 \sin \theta_2)}{(r_2 \cos \theta_2 + ir_2 \sin \theta_2)(r_2 \cos \theta_2 - ir_2 \sin \theta_2)}$$

$$= \frac{r_1r_2(\cos \theta_1 \cos \theta_2 + \sin \theta_1 \sin \theta_2) + ir_1r_2(\sin \theta_1 \cos \theta_2 - \cos \theta_1 \sin \theta_2)}{r_2^2(\cos^2 \theta_2 + \sin^2 \theta_2)}$$

$$= \frac{r_1}{r_2}[\cos(\theta_1 - \theta_2) + i \sin(\theta_1 - \theta_2)]$$

So when you divide two complex numbers you *divide* the modulus of the numerator by the modulus of the denominator and you *subtract* the argument of the denominator from the argument of the numerator.

In the exponential form this means that if $z_1 = r_1 e^{i\theta_1}$ and $z_2 = r_2 e^{i\theta_2}$

- **then** $$z_1 z_2 = r_1 r_2 e^{i(\theta_1 + \theta_2)}$$

- **and** $$\frac{z_1}{z_2} = \frac{r_1}{r_2} e^{i(\theta_1 - \theta_2)}$$

You should learn these results:

- $|z_1 z_2| = |z_1| |z_2|$

- $\left|\dfrac{z_1}{z_2}\right| = \dfrac{|z_1|}{|z_2|}$

- $\arg(z_1 z_2) = \arg z_1 + \arg z_2$

- $\arg \dfrac{z_1}{z_2} = \arg z_1 - \arg z_2$

In the case of the last two results, you may need to add or subtract 2π in order that

$$-\pi < \arg z_1 + \arg z_2 \leqslant \pi$$

and $$-\pi < \arg z_1 - \arg z_2 \leqslant \pi$$

That is, so that you still have the *principal* argument once you have added or subtracted.

Example 4

Given that $$z_1 = 9\left(\cos \tfrac{7\pi}{12} + i \sin \tfrac{7\pi}{12}\right)$$

and $$z_2 = 5\cos\left(\tfrac{2\pi}{3} + i \sin \tfrac{2\pi}{3}\right)$$

find $z_1 z_2$ in the form $a + ib$, $a, b \in \mathbb{R}$.

$$
\begin{aligned}
z_1 z_2 &= 9 \times 5\left[\cos\left(\tfrac{7\pi}{12} + \tfrac{2\pi}{3}\right) + i \sin\left(\tfrac{7\pi}{12} + \tfrac{2\pi}{3}\right)\right] \\
&= 45\left[\cos \tfrac{15\pi}{12} + i \sin \tfrac{15\pi}{12}\right] \\
&= 45\left[\cos \tfrac{5\pi}{4} + i \sin \tfrac{5\pi}{4}\right] \\
&= 45\left[\cos\left(-\tfrac{3\pi}{4}\right) + i \sin\left(-\tfrac{3\pi}{4}\right)\right] \\
&= 45\left[\cos \tfrac{3\pi}{4} - i \sin \tfrac{3\pi}{4}\right] \\
&= 45\left(-\frac{1}{\sqrt{2}} - \frac{i}{\sqrt{2}}\right) = -\frac{45\sqrt{2}}{2}(1 + i)
\end{aligned}
$$

Example 5

Find two values of z for which $\cos z = \frac{5}{4}$.

Using $\cos z = \dfrac{e^{iz} + e^{-iz}}{2}$ you have

$$e^{iz} + e^{-iz} = \tfrac{5}{2} \Rightarrow 2e^{2iz} + 2 = 5e^{iz}$$

Rearranging: $\qquad\qquad 2e^{2iz} - 5e^{iz} + 2 = 0$

Factorising: $\qquad\qquad (2e^{iz} - 1)(e^{iz} - 2) = 0$

That is: $\qquad\qquad e^{iz} = \tfrac{1}{2} \quad \text{or} \quad e^{iz} = 2$

$$iz = -\ln 2 \quad \text{or} \quad \ln 2$$

$$\Rightarrow z = \pm i \ln 2$$

Exercise 2A

1 Write the following in the form (i) $r(\cos\theta + i\sin\theta)$ (ii) $re^{i\theta}$, $-\pi < \theta \leqslant \pi$, giving θ either as a multiple of π or in radians to 3 significant figures.

   (a) $5i$
   (b) 7
   (c) $-3i$
   (d) -6

   (e) $1 + i\sqrt{3}$
   (f) $3\sqrt{3} - 3i$
   (g) $-3 + 4i$
   (h) $1 - i$

   (i) $6 - 8i$
   (j) $\dfrac{2}{1 - i\sqrt{3}}$
   (k) $\dfrac{8}{\sqrt{3} - i}$
   (l) $\dfrac{3 - 2i}{1 + 4i}$

2 Write the following in the form $a + ib$, $a, b \in \mathbb{R}$:

   (a) $3\left(\cos\frac{\pi}{3} + i\sin\frac{\pi}{3}\right)$
   (b) $-5\left(\cos\frac{\pi}{4} - i\sin\frac{\pi}{4}\right)$

   (c) $6\left[\cos\left(-\frac{\pi}{6}\right) + i\sin\left(-\frac{\pi}{6}\right)\right]$
   (d) $-4\left(\cos\frac{3\pi}{2} + i\sin\frac{3\pi}{2}\right)$

   (e) $2\left(\cos\frac{2\pi}{7} + i\sin\frac{2\pi}{7}\right) \times 5\left(\cos\frac{5\pi}{7} + i\sin\frac{5\pi}{7}\right)$

   (f) $\left[3\left(\cos\frac{7\pi}{12} + i\sin\frac{7\pi}{12}\right)\right]^2$
   (g) $\dfrac{7\left(\cos\frac{\pi}{2} + i\sin\frac{\pi}{2}\right)}{3\left(\cos\frac{\pi}{4} + i\sin\frac{\pi}{4}\right)}$

   (h) $\dfrac{6\left(\cos\frac{\pi}{4} + i\sin\frac{\pi}{4}\right)}{2\left(\cos\frac{\pi}{2} - i\sin\frac{\pi}{2}\right)}$
   (i) $\left[2\left(\cos\frac{5\pi}{18} + i\sin\frac{5\pi}{18}\right)\right]^3$

   (j) $\dfrac{\left[2\left(\cos\frac{\pi}{4} + i\sin\frac{\pi}{4}\right)\right]^2}{3\left(\cos\frac{\pi}{3} + i\sin\frac{\pi}{3}\right)}$

3 Simplify, without the use of a calculator

$$\frac{\left(\cos\frac{\pi}{7} - i\sin\frac{\pi}{7}\right)^3}{\left(\cos\frac{\pi}{7} + i\sin\frac{\pi}{7}\right)^4}$$

[E]

4 From the identity

$$\sin(z + w) \equiv \sin z \cos w + \cos z \sin w$$

find the corresponding identity for $\sinh(z + w)$.

5 Assuming the identity for

$$\tan(z + w) \equiv \frac{\tan z + \tan w}{1 - \tan z \tan w}$$

find the corresponding identity for $\tanh(z + w)$.

6 Assuming the identity for

$$\sec^2 z \equiv 1 + \tan^2 z$$

find the corresponding identity for $\operatorname{sech}^2 z$.

7 Use the fact that $e^x \sin 3x = \operatorname{Im}\left(e^x e^{i3x}\right)$ to find

$$\int e^x \sin 3x \, dx$$

8 Use the formula

$$\cos(x + iy) = \cos x \cos iy - \sin x \sin iy$$

to find two imaginary numbers whose cosine is 3.

9 Find the real and imaginary parts of $\tanh(x + iy)$.

10 Find the following in the form $a + ib$, $a, b \in \mathbb{R}$:

(a) $\arccos 4$

(b) $\arcsin 2$

(c) $\arcsin i$

2.4 De Moivre's theorem

In the previous section you found that if

$$z_1 = r_1(\cos\theta_1 + i\sin\theta_1)$$

and $$z_2 = r_2(\cos\theta_2 + i\sin\theta_2)$$

then $$z_1 z_2 = r_1 r_2[\cos(\theta_1 + \theta_2) + i\sin(\theta_1 + \theta_2)]$$

Consequently you should be able to see that if

$$z = r(\cos\theta + i\sin\theta)$$

then
$$z^2 = [r(\cos\theta + i\sin\theta)]^2$$
$$= [r(\cos\theta + i\sin\theta)] \times [r(\cos\theta + i\sin\theta)]$$
$$= r^2[\cos(\theta + \theta) + i\sin(\theta + \theta)]$$
$$= r^2(\cos 2\theta + i\sin 2\theta)$$

Now
$$z^3 = [r(\cos\theta + i\sin\theta)]^3$$
$$= z^2 \times z$$
$$= r^2[\cos 2\theta + i\sin 2\theta] \times r[\cos\theta + i\sin\theta]$$
$$= r^3[\cos(2\theta + \theta) + i\sin(2\theta + \theta)]$$
$$= r^3[\cos 3\theta + i\sin 3\theta]$$

Similarly:
$$z^4 = z^3 \times z$$
$$= r^3[\cos 3\theta + i\sin 3\theta] \times r[\cos\theta + i\sin\theta]$$
$$= r^4(\cos 4\theta + i\sin 4\theta)$$

So the question arises, whether, in general, if $z = r(\cos\theta + i\sin\theta)$ then
$$z^n = r^n(\cos n\theta + i\sin n\theta)$$

Well, if $z = r(\cos\theta + i\sin\theta) = re^{i\theta}$ then:
$$z^n = \left[re^{i\theta}\right]^n = [r(\cos\theta + i\sin\theta)]^n$$

But
$$\left[re^{i\theta}\right]^n = r^n(e^{i\theta})^n = r^n e^{in\theta}$$

using $(a^x)^n = a^{xn}$ from Book P1 (page 1).

Now
$$e^{in\theta} = e^{i(n\theta)} = \cos n\theta + i\sin n\theta$$

So:

■
$$z^n = [r(\cos\theta + i\sin\theta)]^n = r^n(\cos n\theta + i\sin n\theta)$$

This is known as **de Moivre's theorem** and is true for $n \in \mathbb{Q}, \theta \in \mathbb{R}$.

For your syllabus you must be able to prove de Moivre's theorem for any integer n.

Example 6
Express $\cos 5\theta$ in terms of $\cos\theta$.

$(\cos\theta + i\sin\theta)^5 = \cos 5\theta + i\sin 5\theta$, by de Moivre's theorem.

Now, using the binomial theorem (Book C2, page 75) you can expand the left-hand side:

$$(\cos\theta + i\sin\theta)^5 = \cos^5\theta + 5\cos^4\theta(i\sin\theta) + 10\cos^3\theta(i\sin\theta)^2 + 10\cos^2\theta(i\sin\theta)^3$$
$$+ 5\cos\theta(i\sin\theta)^4 + (i\sin\theta)^5$$
$$= \cos^5\theta + 5i\cos^4\theta\sin\theta - 10\cos^3\theta\sin^2\theta - 10i\cos^2\theta\sin^3\theta$$
$$+ 5\cos\theta\sin^4\theta + i\sin^5\theta$$

If you consider only the real parts, you have:

$$\cos 5\theta = \cos^5\theta - 10\cos^3\theta\sin^2\theta + 5\cos\theta\sin^4\theta$$

But
$$\sin^2\theta \equiv 1 - \cos^2\theta$$

So:

$$\cos 5\theta = \cos^5\theta - 10\cos^3\theta(1 - \cos^2\theta) + 5\cos\theta(1 - \cos^2\theta)^2$$
$$= \cos^5\theta - 10\cos^3\theta + 10\cos^5\theta + 5\cos\theta(1 - 2\cos^2\theta + \cos^4\theta)$$
$$= \cos^5\theta - 10\cos^3\theta + 10\cos^5\theta + 5\cos\theta - 10\cos^3\theta + 5\cos^5\theta$$
$$\cos 5\theta = 16\cos^5\theta - 20\cos^3\theta + 5\cos\theta$$

Example 7

Evaluate $(\cos\frac{\pi}{6} + i\sin\frac{\pi}{6})^{-3}$.

$$(\cos\tfrac{\pi}{6} + i\sin\tfrac{\pi}{6})^{-3} = \cos\left(-\tfrac{3\pi}{6}\right) + i\sin\left(-\tfrac{3\pi}{6}\right)$$
$$= \cos\left(-\tfrac{\pi}{2}\right) + i\sin\left(-\tfrac{\pi}{2}\right)$$
$$= \cos\tfrac{\pi}{2} - i\sin\tfrac{\pi}{2}$$
$$= -i$$

Example 8

Simplify
$$\frac{\left(\cos\frac{5\pi}{11} + i\sin\frac{5\pi}{11}\right)^7}{\left(\cos\frac{3\pi}{11} - i\sin\frac{3\pi}{11}\right)^3}$$

$$\frac{\left(\cos\frac{5\pi}{11} + i\sin\frac{5\pi}{11}\right)^7}{\left(\cos\frac{3\pi}{11} - i\sin\frac{3\pi}{11}\right)^3} = \frac{\cos\frac{35\pi}{11} + i\sin\frac{35\pi}{11}}{\left[\cos\left(-\frac{3\pi}{11}\right) + i\sin\left(-\frac{3\pi}{11}\right)\right]^3}$$

$$= \frac{\cos\frac{35\pi}{11} + i\sin\frac{35\pi}{11}}{\cos\left(-\frac{9\pi}{11}\right) + i\sin\left(-\frac{9\pi}{11}\right)}$$

$$= \cos\left(\tfrac{35\pi}{11} + \tfrac{9\pi}{11}\right) + i\sin\left(\tfrac{35\pi}{11} + \tfrac{9\pi}{11}\right)$$

$$= \cos 4\pi + i\sin 4\pi$$

$$= 1$$

Example 9

Express $\cos^4\theta$ in terms of $\cos 4\theta$ and $\cos 2\theta$.

If $z = \cos\theta + i\sin\theta$, then:

$$\frac{1}{z} = z^{-1} = (\cos\theta + i\sin\theta)^{-1}$$
$$= \cos(-\theta) + i\sin(-\theta)$$
$$= \cos\theta - i\sin\theta$$

So:
$$z + \frac{1}{z} = 2\cos\theta$$

Similarly:
$$z^n = \cos n\theta + i\sin n\theta$$

and
$$\frac{1}{z^n} = z^{-n} = \cos(-n\theta) + i\sin(-n\theta)$$
$$= \cos n\theta - i\sin n\theta$$

So:
$$z^n + \frac{1}{z^n} = 2\cos n\theta$$

Now
$$\left(z + \frac{1}{z}\right)^4 = z^4 + 4z^3\left(\frac{1}{z}\right) + 6z^2\left(\frac{1}{z}\right)^2 + 4z\left(\frac{1}{z}\right)^3 + \left(\frac{1}{z}\right)^4$$
$$= z^4 + 4z^2 + 6 + \frac{4}{z^2} + \frac{1}{z^4}$$
$$= \left(z^4 + \frac{1}{z^4}\right) + 4\left(z^2 + \frac{1}{z^2}\right) + 6$$
$$= 2\cos 4\theta + 4(2\cos 2\theta) + 6$$
$$= 2\cos 4\theta + 8\cos 2\theta + 6$$

But
$$\left(z + \frac{1}{z}\right)^4 = (2\cos\theta)^4 = 16\cos^4\theta$$

So
$$\cos^4\theta = \tfrac{1}{16}(2\cos 4\theta + 8\cos 2\theta + 6)$$
$$= \tfrac{1}{8}\cos 4\theta + \tfrac{1}{2}\cos 2\theta + \tfrac{3}{8}$$

2.5 The cube roots of unity

If you want to find the cube roots of 1 then you let
$$r(\cos\theta + i\sin\theta) = 1$$

But 1 has modulus 1, so $r = 1$.

Thus:
$$\cos\theta + i\sin\theta = 1$$

So
$$\cos\theta = 1 \quad \text{and} \quad \sin\theta = 0$$

That is:
$$\theta = 0 \,(\text{principal value}),\ 2\pi,\ 4\pi,\ \text{etc.}$$

This means
$$\sqrt[3]{1} = (\cos 0 + i \sin 0)^{\frac{1}{3}} = \cos \frac{0}{3} + i \sin \frac{0}{3}$$

by de Moivre's theorem.

So:
$$\sqrt[3]{1} = 1$$

Also
$$\sqrt[3]{1} = (\cos 2\pi + i \sin 2\pi)^{\frac{1}{3}}$$
$$= \cos \tfrac{2\pi}{3} + i \sin \tfrac{2\pi}{3}$$

and
$$\sqrt[3]{1} = (\cos 4\pi + i \sin 4\pi)^{\frac{1}{3}}$$
$$= \cos \tfrac{4\pi}{3} + i \sin \tfrac{4\pi}{3}$$
$$= \cos \left(-\tfrac{2\pi}{3}\right) + i \sin \left(-\tfrac{2\pi}{3}\right)$$

Any number has only three cube roots and to check you can take the next value of $\theta = 6\pi$:
$$\sqrt[3]{1} = (\cos 6\pi + i \sin 6\pi)^{\frac{1}{3}}$$
$$= \cos 2\pi + i \sin 2\pi = 1, \text{ again}$$

So as you now take values of $\theta = 6\pi,\ 8\pi,\ 10\pi,$ etc., you only get the same three roots repeated again and again.

If you plot these three cube roots on an Argand diagram they look like this:

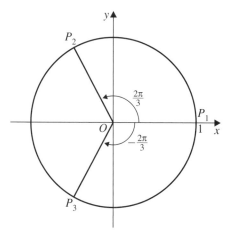

Notice the following points:

(i) You can see that the points P_1, P_2, P_3 representing the roots lie on a circle, centre the origin and radius 1 (because $r = 1$) and that the angle between each consecutive point is the same: $\frac{2\pi}{3}$. That is:
$$\angle P_1 O P_2 = \angle P_2 O P_3 = \angle P_3 O P_1 = \tfrac{2\pi}{3}$$

(ii) If
$$\cos \tfrac{2\pi}{3} + i \sin \tfrac{2\pi}{3} = \omega$$

then:
$$\cos \tfrac{4\pi}{3} + i \sin \tfrac{4\pi}{3} = \left(\cos \tfrac{2\pi}{3} + i \sin \tfrac{2\pi}{3}\right)^2 = \omega^2$$

So the cube roots of unity can be written as $1,\ \omega,\ \omega^2$.

(iii) Now
$$\omega^2 = \cos \tfrac{4\pi}{3} + i \sin \tfrac{4\pi}{3}$$
$$= \cos\left(-\tfrac{2\pi}{3}\right) + i \sin\left(-\tfrac{2\pi}{3}\right)$$
$$= \cos \tfrac{2\pi}{3} - i \sin \tfrac{2\pi}{3}$$
$$= \left(\cos \tfrac{2\pi}{3} + i \sin \tfrac{2\pi}{3}\right)^* = \omega^*$$

That is, ω is the *conjugate* of ω^2 and vice versa. (Book FP1, page 25.)

(iv) You can see a fourth property of these cube roots if you consider

$$1 + \omega + \omega^2 = 1 + \left(\cos \tfrac{2\pi}{3} + i \sin \tfrac{2\pi}{3}\right) + \left(\cos \tfrac{2\pi}{3} - i \sin \tfrac{2\pi}{3}\right)$$
$$= 1 + 2\cos \tfrac{2\pi}{3}$$

But
$$\cos \tfrac{2\pi}{3} = -\cos \tfrac{\pi}{3} = -\tfrac{1}{2}$$

So
$$1 + \omega + \omega^2 = 1 + 2\left(-\tfrac{1}{2}\right) = 0$$

2.6 The *n*th roots of unity

If you now extend this to find the *n*th roots of unity, where $n \in \mathbb{Z}^+$, then

$$\cos \theta + i \sin \theta = 1$$
$$\Rightarrow \quad \theta = 0,\, 2\pi,\, 4\pi,\, 6\pi,\, \text{etc.}$$

A first *n*th root is given by

$$(\cos 0 + i \sin 0)^{\frac{1}{n}} = \cos 0 + i \sin 0 = 1$$

A second root is given by

$$(\cos 2\pi + i \sin 2\pi)^{\frac{1}{n}} = \cos \frac{2\pi}{n} + i \sin \frac{2\pi}{n}$$

A third root is given by

$$\cos \frac{4\pi}{n} + i \sin \frac{4\pi}{n}$$

The next is

$$\cos \frac{6\pi}{n} + i \sin \frac{6\pi}{n}$$

and so on.

On an Argand diagram they look like this:

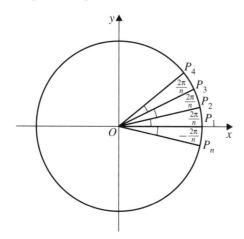

The properties of the *n*th roots of unity are:

■ **One root is always 1.**

■ **If *n* is even, another root is −1.**

■ **There are *n* roots and they are represented on an Argand diagram by points on a unit circle where**

$$\angle P_1OP_2 = \angle P_2OP_3 = \ldots = \angle P_nOP_1 = \frac{2\pi}{n}$$

If the root represented by P_2 is ω, that is:

$$\omega = \cos\frac{2\pi}{n} + \text{i}\sin\frac{2\pi}{n}$$

then the root represented by P_3 is

$$\cos\frac{4\pi}{n} + \text{i}\sin\frac{4\pi}{n} = \left(\cos\frac{2\pi}{n} + \text{i}\sin\frac{2\pi}{n}\right)^2 = \omega^2$$

and the root represented by P_4 is

$$\cos\frac{6\pi}{n} + \text{i}\sin\frac{6\pi}{n} = \left(\cos\frac{2\pi}{n} + \text{i}\sin\frac{2\pi}{n}\right)^3 = \omega^3$$

■ **The *n*th roots of unity can be written as**

$$1, \omega, \omega^2, \omega^3, \ldots, \omega^{n-2}, \omega^{n-1}$$

$$\omega = \cos\frac{2\pi}{n} + \text{i}\sin\frac{2\pi}{n}$$

and $\omega^{n-1} = \cos\left(-\frac{2\pi}{n}\right) + \text{i}\sin\left(-\frac{2\pi}{n}\right) = \cos\frac{2\pi}{n} - \text{i}\sin\frac{2\pi}{n} = \omega^*$

$$\omega^2 = \cos\frac{4\pi}{n} + \text{i}\sin\frac{4\pi}{n}$$

and

$$\omega^{n-2} = \cos\left(-\frac{4\pi}{n}\right) + i\sin\left(-\frac{4\pi}{n}\right) = \cos\frac{4\pi}{n} - i\sin\frac{4\pi}{n} = \left(\omega^2\right)^*$$

■ **So, with the exception of 1 (and −1 if *n* is even) the other roots occur in conjugate pairs.**

Finally, $\qquad 1 + \omega + \omega^2 + \ldots + \omega^{n-2} + \omega^{n-1} = \dfrac{1-\omega^n}{1-\omega}$

using the formula for the sum of a geometric series (Book P1, page 120).

But since ω is an *n*th root of 1 then

$$\omega^n = 1$$

So: $\qquad 1 + \omega + \omega^2 + \ldots + \omega^{n-2} + \omega^{n-1} = \dfrac{1-1}{1-\omega} = 0$

■ **So the sum of the *n*th roots of unity is zero.**

2.7 The *n*th roots of any complex number

If you have a complex number

$$z = r(\cos\theta + i\sin\theta)$$

then z can also be written:

$$z = r[\cos(\theta + 2k\pi) + i\sin(\theta + 2k\pi)], \, k \in \mathbb{Z}$$

because adding (or subtracting) multiples of 2π to the principal argument takes you to the same point where you started on an Argand diagram.

The *n*th roots of z are then given by

$$z^{\frac{1}{n}} = \{r[\cos(\theta + 2k\pi) + i\sin(\theta + 2k\pi)]\}^{\frac{1}{n}}$$

$$= r^{\frac{1}{n}}\left[\cos\left(\frac{\theta + 2k\pi}{n}\right) + i\sin\left(\frac{\theta + 2k\pi}{n}\right)\right]$$

By letting $k = 0, 1, 2, \ldots, n-1$, you can obtain the *n*th roots of z. In exponential form these will be

$$r^{\frac{1}{n}}e^{\frac{i\theta}{n}}, \, r^{\frac{1}{n}}e^{i\left(\frac{\theta+2\pi}{n}\right)}, \, r^{\frac{1}{n}}e^{i\left(\frac{\theta+4\pi}{n}\right)}, \ldots, r^{\frac{1}{n}}e^{i\left[\frac{\theta+2(n-1)\pi}{n}\right]}$$

Example 10
Find the fifth roots of $1 + i$.

From the diagram you can see that if P represents $1 + i$ then

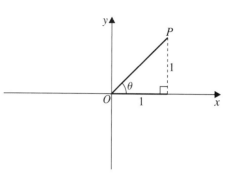

$$OP = \sqrt{2} \quad \text{and} \quad \theta = \tfrac{\pi}{4}$$

So

$$1 + i = \sqrt{2}\left[\cos\tfrac{\pi}{4} + i\sin\tfrac{\pi}{4}\right]$$

$$= \sqrt{2}\left[\cos\left(\tfrac{\pi}{4} + 2k\pi\right) + i\sin\left(\tfrac{\pi}{4} + 2k\pi\right)\right], k \in \mathbb{Z}$$

$$(1+i)^{\frac{1}{5}} = \left(\sqrt{2}\right)^{\frac{1}{5}}\left[\cos\left(\frac{\tfrac{\pi}{4} + 2k\pi}{5}\right) + i\sin\left(\frac{\tfrac{\pi}{4} + 2k\pi}{5}\right)\right]$$

If $k = 0$ then a root is $\left(\sqrt{2}\right)^{\frac{1}{5}}\left[\cos\tfrac{\pi}{20} + i\sin\tfrac{\pi}{20}\right]$

If $k = 1$ then another root is $\left(\sqrt{2}\right)^{\frac{1}{5}}\left[\cos\tfrac{9\pi}{20} + i\sin\tfrac{9\pi}{20}\right]$

If $k = 2$ a third root is $\left(\sqrt{2}\right)^{\frac{1}{5}}\left[\cos\tfrac{17\pi}{20} + i\sin\tfrac{17\pi}{20}\right]$

If $k = 3$ a fourth root is

$$\left(\sqrt{2}\right)^{\frac{1}{5}}\left[\cos\tfrac{25\pi}{20} + i\sin\tfrac{25\pi}{20}\right] = \left(\sqrt{2}\right)^{\frac{1}{5}}\left[\cos\left(\tfrac{-15\pi}{20}\right) + i\sin\left(\tfrac{-15\pi}{20}\right)\right] = \left(\sqrt{2}\right)^{\frac{1}{5}}\left(\cos\tfrac{3\pi}{4} - i\sin\tfrac{3\pi}{4}\right)$$

If $k = 4$ a fifth root is

$$\left(\sqrt{2}\right)^{\frac{1}{5}}\left[\cos\tfrac{33\pi}{20} + i\sin\tfrac{33\pi}{20}\right] = \left(\sqrt{2}\right)^{\frac{1}{5}}\left[\cos\left(\tfrac{-7\pi}{20}\right) + i\sin\left(\tfrac{-7\pi}{20}\right)\right] = \left(\sqrt{2}\right)^{\frac{1}{5}}\left(\cos\tfrac{7\pi}{20} - i\sin\tfrac{7\pi}{20}\right)$$

Exercise 2B

1 Use de Moivre's theorem to simplify:

(a) $\left[\cos\tfrac{2\pi}{5} + i\sin\tfrac{2\pi}{5}\right]^{10}$ (b) $\left[\cos\tfrac{\pi}{12} + i\sin\tfrac{\pi}{12}\right]^{8}$

(c) $\left[\cos\left(-\tfrac{\pi}{27}\right) + i\sin\left(-\tfrac{\pi}{27}\right)\right]^{9}$ (d) $\left(\cos\tfrac{\pi}{18} - i\sin\tfrac{\pi}{18}\right)^{3}$

2 Express $z = 2(1 - i\sqrt{3})$ in the form $r(\cos\theta + i\sin\theta)$.
Hence find z^{8} and $\dfrac{1}{z^{5}}$ in the form $a + ib$.

3 Express $z = (1 - i)$ in the form $r(\cos\theta + i\sin\theta)$. Hence find z^{4}
and $\dfrac{1}{z^{7}}$ in the form $a + ib$.

4 Simplify $\dfrac{(\cos\tfrac{\pi}{9} + i\sin\tfrac{\pi}{9})^{4}}{(\cos\tfrac{\pi}{9} - i\sin\tfrac{\pi}{9})^{5}}$ [E]

5 Find $\sin 5\theta$ in terms of $\sin\theta$.

6 Find $\sin 3\theta$ in terms of $\sin\theta$.

7 Find $\cos 7\theta$ in terms of $\cos\theta$.

8 Find $\sin 7\theta$ in terms of $\sin\theta$.

9 Find $\tan 3\theta$ in terms of $\tan\theta$.

10 Find $\tan 5\theta$ in terms of $\tan\theta$.

11 Express (a) $\cos^5\theta$ (b) $\cos^6\theta$ (c) $\cos^7\theta$ in terms of cosines of multiples of θ.

12 Express (a) $\sin^4\theta$ in terms of cosines of multiples of θ
(b) (i) $\sin^5\theta$ (ii) $\sin^7\theta$ in terms of sines of multiples of θ.

13 Find (a) $\displaystyle\int \sin^4\theta\,d\theta$ (b) $\displaystyle\int \cos^6\theta\,d\theta$ (c) $\displaystyle\int \sin^4\theta\cos^2\theta\,d\theta$.

14 Find, in the form $re^{i\theta}$, the cube roots of:
(a) i (b) -1 (c) $-5+12i$ (d) $\dfrac{1+i}{1-i}$

15 Solve the equation
$$z^5 + 1 = 0$$

16 Find the cube roots of $21 + 72i$.

17 Find the fourth roots of unity.

18 If $2\cos\theta = z + z^{-1}$, prove that, if n is a positive integer,
$$2\cos n\theta = z^n + z^{-n}$$
Hence, or otherwise, solve the equation
$$3z^4 - z^3 + 2z^2 - z + 3 = 0$$
given that no root is real. [E]

19 If $z = \cos\theta + i\sin\theta$, show that
$$z^n + z^{-n} = 2\cos n\theta \quad\text{and}\quad z^n - z^{-n} = 2i\sin n\theta$$
Hence deduce that
$$\cos^6\theta + \sin^6\theta = \tfrac{1}{8}(3\cos 4\theta + 5)$$ [E]

20 Express $\sqrt{3} - i$ in the form $r(\cos\theta + i\sin\theta)$. Hence show that $\left(\sqrt{3} - i\right)^9$ can be expressed as ci, where c is real, and give the value of c. [E]

21 Find the roots of the equation $(z+1)^3 = 1$ in the form $x + iy$ and show points representing these roots on an Argand diagram. [E]

22 Use de Moivre's theorem to prove that, if θ is not a multiple of π,
$$\frac{\sin 5\theta}{\sin\theta} = 16\cos^4\theta - 12\cos^2\theta + 1$$ [E]

23 Prove that
$$(z^n - e^{i\theta})(z^n - e^{-i\theta}) = z^{2n} - 2z^n \cos\theta + 1$$
Hence, or otherwise, find the roots of the equation
$$z^6 - z^3\sqrt{2} + 1 = 0$$
in the form $\cos\alpha + i\sin\alpha$, $-\pi < \alpha \leqslant \pi$. [E]

24 Find, in the form $re^{i\theta}$, $r > 0$ and $-\pi < \theta \leqslant \pi$, the four fourth roots of $8(-1 + i\sqrt{3})$, and plot them on an Argand diagram. [E]

2.8 Loci in the complex plane

As you know, you can represent complex numbers on an Argand diagram. If you place some restrictions on the complex number, for example $|z| = 5$ or $\arg z = \frac{\pi}{3}$, then z cannot be *any* complex number – it can only take values that satisfy the conditions laid down. Consequently, when you come to represent these values of z which satisfy the conditions on the Argand diagram you find that the various possible positions of z form a curve or a straight line. This path, which represents all possible values of z, is called the **locus** of z.

Most of the equations that you will meet in this section will involve either the modulus or the argument of complex numbers. If you have an equation involving the argument you need to remember that the argument is the angle which the vector representing the number on the Argand diagram makes with the positive x-axis. You also need to remember, and use, the fact that

$$\arg \frac{z_1}{z_2} = \arg z_1 - \arg z_2$$

When you have an equation involving a modulus you must first of all remember what the representation of $z - a$ looks like on an Argand diagram:

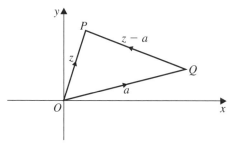

If $\overrightarrow{OP}$ represents the number z and if $\overrightarrow{OQ}$ represents the number a then $\overrightarrow{QP}$ represents the number $z - a$. Consequently the length of the line $QP = |\overrightarrow{QP}|$ represents $|z - a|$.

Example 11
Sketch the locus of z when $|z| = 5$.

If $z = x + iy$ then z is represented on the Argand diagram by the vector going from the origin to the point P with coordinates (x, y). So the equation $|z| = 5$ states that the point P can only be in positions such that the length of the vector $\overrightarrow{OP}$ is 5 units. Thus the locus is a circle, centre the origin, with radius 5 units.

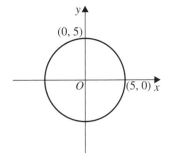

Example 12
Sketch the locus of z given that $|z - 2 + 5i| = 3$.

$$z - 2 + 5i = z - (2 - 5i)$$

So the vector representing $z - (2 - 5i)$ starts at $(2, -5)$ and goes to (x, y). The equation $|z - 2 + 5i| = 3$ thus states that the length of the vector that starts at $(2, -5)$ and goes to (x, y) is 3 units. That is, the locus of z is a circle with centre at $(2, -5)$ and with radius 3.

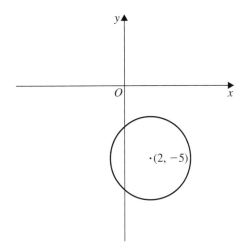

Example 13
Sketch the locus of z when $\arg z = \frac{\pi}{4}$.

If $\arg z = \frac{\pi}{4}$ then z can be any complex number such that the vector z which goes from the origin on the Argand diagram to the point (x, y) makes an angle of $\frac{\pi}{4}$ with the positive x-axis. So the locus of z is a **half-line**.

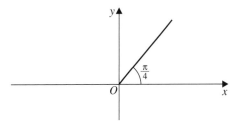

Example 14

Sketch the locus of z such that $\arg(z+3) = -\frac{\pi}{6}$.

$$z + 3 = z - (-3)$$

The vector representing $z + 3$ starts at $(-3, 0)$ and goes to (x, y).
So this vector makes an angle of $-\frac{\pi}{6}$ with the positive x-axis.
The locus of z is another half-line.

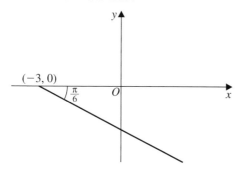

Example 15

Sketch the locus of z such that $|z + 2| = |z - \mathrm{i}|$.

You should by now be able to see that, on an Argand diagram,
$|z + 2|$ is the length of the line going from $(-2, 0)$ to (x, y) and
$|z - \mathrm{i}|$ is the length of the line going from $(0, 1)$ to (x, y). So the
equation $|z + 2| = |z - \mathrm{i}|$ states that z can be represented by any
point $P(x, y)$ such that the distance of the point P from $(-2, 0)$ is
equal to its distance from $(0, 1)$. The locus of z is the perpendicular
bisector of the line joining $(-2, 0)$ to $(0, 1)$.

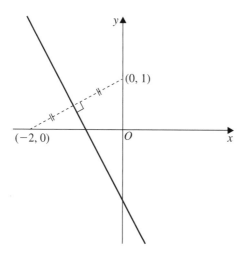

Example 16

Sketch the locus of z such that

$$|z - 3| = 2|z - 1 + \mathrm{i}|$$

This means that the locus is the path of the point (x, y) such that
the distance between $(3, 0)$ and (x, y) is twice the distance between

$(1, -1)$ and (x, y). It is all very well to state this, but what sort of a curve does this give you? Since on this occasion the locus is not obvious, you need to resort to algebra. (You could have done this with the previous examples if you had got stuck.)

If $z = x + iy$ then:

$$|z - 3| = 2|z - 1 + i|$$
$$\Rightarrow \quad |x + iy - 3| = 2|x + iy - 1 + i|$$
$$\Rightarrow \quad |(x - 3) + iy| = 2|(x - 1) + i(y + 1)|$$
$$\Rightarrow \quad (x - 3)^2 + y^2 = 4\left[(x - 1)^2 + (y + 1)^2\right]$$
$$x^2 - 6x + 9 + y^2 = 4(x^2 - 2x + 1 + y^2 + 2y + 1)$$
$$x^2 + y^2 - 6x + 9 = 4x^2 + 4y^2 - 8x + 8y + 8$$

So:

$$3x^2 + 3y^2 - 2x + 8y - 1 = 0$$
$$\Rightarrow x^2 + y^2 - \tfrac{2}{3}x + \tfrac{8}{3}y = \tfrac{1}{3}$$
$$\left(x - \tfrac{1}{3}\right)^2 + \left(y + \tfrac{4}{3}\right)^2 = \tfrac{1}{3} + \tfrac{1}{9} + \tfrac{16}{9}$$
$$\left(x - \tfrac{1}{3}\right)^2 + \left(y + \tfrac{4}{3}\right)^2 = \tfrac{20}{9}$$

This you should recognise from the work you did in Book C2 as the equation of a circle, centre $\left(\tfrac{1}{3}, -\tfrac{4}{3}\right)$, radius $\tfrac{\sqrt{20}}{3}$.

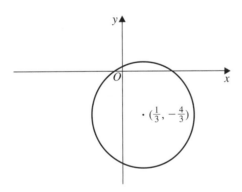

Example 17

Sketch the locus of z such that $\arg \dfrac{z - 2}{z + 5} = \tfrac{\pi}{4}$.

$z - 2$ is represented by a vector from $(2, 0)$ to (x, y).

$z + 5$ is represented by a vector from $(-5, 0)$ to (x, y).

Since $\qquad \arg \dfrac{z - 2}{z + 5} = \arg(z - 2) - \arg(z + 5)$

the equation $\qquad \arg \dfrac{z - 2}{z + 5} = \tfrac{\pi}{4}$

states that when you subtract the angle which the vector representing $z + 5$ makes with the positive x-axis from the angle which the vector representing $z - 2$ makes with the positive x-axis, you get $\frac{\pi}{4}$. The situation looks like this:

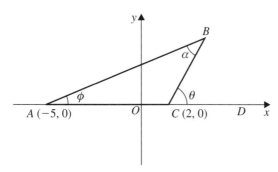

That is:
$$\theta - \phi = \tfrac{\pi}{4}$$

Now, by the geometry of the triangle, you know that
$$\angle ABC + \angle BAC = \angle BCD$$

That is:
$$\alpha + \phi = \theta$$

or
$$\alpha = \theta - \phi$$

So:
$$\angle ABC = \tfrac{\pi}{4}$$

Again using geometry, you should recognise that if B can vary such that $\angle ABC = \frac{\pi}{4}$ then B lies on an arc of a circle passing through A and C.

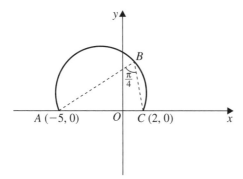

Exercise 2C

1 Sketch the locus of z if:

(a) $|z| = 4$

(b) $|z| = 7$

(c) $|z - i| = 3$

(d) $|z + 3| = 2$

(e) $|z - 1 + 2i| = 5$

(f) $|z + 1 + 3i| = 4$

(g) $|z + 2 - 4i| = 5$

(h) $|2z + 3 - 2i| = 4$

2 Sketch the locus of z if:

(a) $\arg z = \frac{\pi}{6}$

(b) $\arg z = -\frac{\pi}{4}$

(c) $\arg z = -\frac{3\pi}{4}$

(d) $\arg(z+1) = -\frac{\pi}{3}$

(e) $\arg(z-i) = \frac{2\pi}{3}$

(f) $\arg(z+1-i) = \pi$

(g) $\arg(z-1+2i) = \frac{\pi}{3}$

(h) $\arg(z-3+2i) = \frac{\pi}{2}$

3 Sketch the locus of z where:

(a) $|z+i| = |z-1|$

(b) $|z-2i| = |z|$

(c) $|z+1| = |z-2|$

(d) $\dfrac{|z+1-i|}{|z+2+i|} = 1$

(e) $\dfrac{|z-3+2i|}{|z+2-3i|} = 1$

4 Sketch the locus of z where:

(a) $|z+1| = 2|z-1|$

(b) $|z+3i| = 3|z-5|$

(c) $\dfrac{|z+i|}{|z-3i|} = 3$

(d) $\dfrac{z}{|z-2i|} = 5$

(e) $\dfrac{|z+1-3i|}{|z-2+i|} = 3$

5 Sketch the locus of z where:

(a) $\arg\left(\dfrac{z-1}{z+1}\right) = \frac{\pi}{4}$

(b) $\arg\left(\dfrac{z-i}{z+2}\right) = \frac{\pi}{3}$

(c) $\arg\left(\dfrac{z-2i}{z-3}\right) = \frac{\pi}{6}$

(d) $\arg\left(\dfrac{z}{z-3i}\right) = \frac{\pi}{2}$

(e) $\arg\left(\dfrac{z}{z-1+2i}\right) = \frac{\pi}{3}$

6 Represent on an Argand diagram the loci $|z-2| = 2$ and $|z-4| = 2$. Calculate the complex numbers corresponding to the points of intersection of these loci. [E]

7 Sketch the curve in the Argand diagram defined by $|z-1| = 1, \operatorname{Im} z \geqslant 0$. Find the value of z at the point P in which this curve is cut by the line $|z-1| = |z-2|$. Find also the value of $\arg z$ and $\arg(z-2)$ at P. [E]

8 Draw the line $|z| = |z-4|$ and the half-line $\arg(z-i) = \frac{\pi}{4}$ in the Argand diagram. Hence find the complex number which satisfies both equations. [E]

9 In an Argand diagram the point A has coordinates $(1, 0)$ and the point B has coordinates $(0, 2)$. The point P represents the complex number z. Given that

$$\arg\left(\frac{z-1}{z-2i}\right) = \frac{\pi}{4}$$

describe the locus of P and sketch this locus on an Argand diagram. Show that the point $(1, 3)$ lies on this locus.   [E]

10 If $z = x + iy$, find the real and imaginary parts of $z + \frac{1}{z}$.

Find the locus of points in an Argand diagram for which the imaginary part of $z + \frac{1}{z}$ is zero.   [E]

11 If $|z - 1| = 3|z + 1|$, prove that the locus of z is a circle and find its centre and radius.   [E]

12 Show that in an Argand diagram the equation

$$\arg(z-2) - \arg(z-2i) = \frac{3\pi}{4}$$

represents an arc of a circle and that $\dfrac{|z-4|}{|z-1|}$ is constant on this circle.

Find the values of z corresponding to the points in which this circle is cut by the curve given by

$$|z-1| + |z-4| = 5$$ [E]

13 Sketch on an Argand diagram the curve described by the equation

$$|z - 3 + 6i| = 2|z|$$

and express the equation of the curve in cartesian form.   [E]

14 Given that

$$\arg\left(\frac{z-1}{z+1}\right) = \frac{\pi}{4}$$

(a) show that the point P, which represents z on an Argand diagram, lies on an arc of a circle.

(b) Find the centre and radius of this circle and sketch the locus of P.

(c) Sketch, on a separate Argand diagram, the locus represented by

$$\left|\frac{z - 3i}{z + 3}\right| = 1$$ [E]

2.9 Modulus inequalities

Let the vector $\overrightarrow{OP}$ represent the complex number z_1.

Let the vector $\overrightarrow{PQ}$ represent the complex number z_2.

Then $\overrightarrow{OQ}$ represents $z_1 + z_2$.

So the distances OP, PQ and OQ represent $|z_1|$, $|z_2|$ and $|z_1 + z_2|$ respectively.

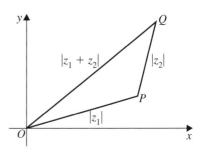

Now you can see that in the $\triangle OPQ$

$$OQ \leqslant OP + PQ$$

and that $OQ = OP + PQ$ only if P lies on the line OQ.

- **Thus:** $\qquad |z_1 + z_2| \leqslant |z_1| + |z_2|$

Similarly, by the geometry of $\triangle OPQ$:

$$OP \leqslant OQ + PQ$$

That is: $|z_1| \leqslant |z_1 + z_2| + |z_2|$

- **or** $\qquad |z_1 + z_2| \geqslant |z_1| - |z_2|$

By interchanging z_1 and z_2 you get

- $\qquad |z_1 + z_2| \geqslant |z_2| - |z_1|$

If you put these last two inequalities together you get

- $\qquad |z_1 + z_2| \geqslant ||z_1| - |z_2||$

These results are called the **triangle inequalities**.

Example 18

Shade on an Argand diagram the region that represents

$$|z - 2| \leqslant 3$$

The equation $|z - 2| = 3$ is represented on an Argand diagram by a circle, centre $(2, 0)$, radius 3. So the inequality $|z - 2| < 3$ is represented by the inside of this circle.

$|z - 2| \leqslant 3$ is thus represented by the inside *and* the boundary of the circle.

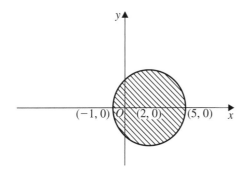

Example 19

Shade the region on an Argand diagram that represents

$$|z - 3| < |z - \mathrm{i}|$$

The equation $|z - 3| = |z - \mathrm{i}|$ is represented on an Argand diagram by the perpendicular bisector of the line joining $(3, 0)$ to $(0, 1)$. So $|z - 3| < |z - \mathrm{i}|$ is that side of the bisector which is closer to $(3, 0)$ than to $(0, 1)$.

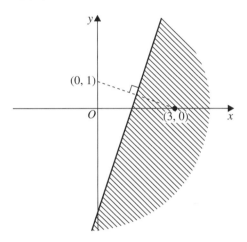

2.10 Elementary transformations from the *z*-plane to the *w*-plane

Suppose two complex variables z and w are connected by a simple relation such as $w = z^2$ or $w = \dfrac{az + b}{cz + d}$, where $a, b, c, d \in \mathbb{C}$. Let the point P represent z in one Argand diagram, which is called the *z*-plane, and let P' represent w in another Argand diagram, which is called the *w*-plane. As z varies so w will vary according to the relation between them. So P and P' will take differing positions in their respective planes. If P describes a simple locus in the *z*-plane then P' will describe a related locus in the *w*-plane. Here are some examples.

Example 20

For the transformation

$$w = \frac{2}{\mathrm{i} + 2z}, \quad z \neq -\frac{\mathrm{i}}{2}$$

find the locus of w as z moves on the circle with equation $|z| = 1$.

$$w = \frac{2}{\mathrm{i} + 2z}$$

$$\Rightarrow \quad wi + 2wz = 2$$

$$2wz = 2 - iw$$

$$z = \frac{2 - iw}{2w}$$

$$\Rightarrow \quad |z| = \left| \frac{2 - iw}{2w} \right|$$

Since $|z| = 1$:

$$1 = \left| \frac{2 - iw}{2w} \right|$$

So:

$$|2w| = |2 - iw|$$

Let $w = u + iv$

Then:

$$|2u + 2iv| = |2 - i(u + iv)|$$

$$\sqrt{(4u^2 + 4v^2)} = |(2 + v) - iu| = \sqrt{\left[(2 + v)^2 + u^2 \right]}$$

$$4u^2 + 4v^2 = (2 + v)^2 + u^2$$

$$3u^2 + 3v^2 - 4v = 4$$

$$u^2 + v^2 - \tfrac{4}{3}v = \tfrac{4}{3}$$

$$u^2 + \left(v - \tfrac{2}{3} \right)^2 = \tfrac{4}{3} + \tfrac{4}{9} = \tfrac{16}{9}$$

So the locus of w is a circle, centre $\left(0, \tfrac{2}{3} \right)$, radius $\tfrac{4}{3}$.

Example 21

Find the locus of w given that $w = z + \dfrac{1}{z}$ and that z moves on the circle with equation $|z| = 3$.

Let $z = re^{i\theta}$; then

$$\frac{1}{z} = \frac{1}{r} e^{-i\theta} = \frac{1}{r} (\cos \theta - i \sin \theta)$$

But $|z| = 3$, so $r = 3$.

So:

$$w = u + iv = \left(r \cos \theta + \frac{1}{r} \cos \theta \right) + i \left(r \sin \theta - \frac{1}{r} \sin \theta \right)$$

$$\Rightarrow \quad u = \tfrac{10}{3} \cos \theta \quad \text{and} \quad v = \tfrac{8}{3} \sin \theta$$

$$\cos^2 \theta + \sin^2 \theta \equiv 1 \Rightarrow \left(\frac{3u}{10} \right)^2 + \left(\frac{3v}{8} \right)^2 = 1$$

That is:

$$\frac{9u^2}{100} + \frac{9v^2}{64} = 1$$

which you should recognise as the equation of an ellipse (see Book FP2).

Example 22

The transformation

$$w = \frac{z+i}{iz+2}, \quad z \neq 2i$$

maps the complex number $z = x + iy$ onto the complex number $w = u + iv$. Find the two points in the complex plane which are invariant under the transformation (that is, the two points where $w = z$).

Show that if z lies on the real axis, then w lies on the circle with equation $2u^2 + 2v^2 + v - 1 = 0$.

$$w = \frac{z+i}{iz+2}$$

If z is invariant then

$$z = \frac{z+i}{iz+2}$$

$$\Rightarrow \quad iz^2 + 2z - z - i = 0$$

$$iz^2 + z - i = 0$$

$$z^2 - iz - 1 = 0$$

So:

$$z = \frac{i \pm \sqrt{[(-i)^2 + 4]}}{2}$$

$$= \frac{i \pm \sqrt{3}}{2}$$

So the invariant points are $\dfrac{i + \sqrt{3}}{2}$ and $\dfrac{i - \sqrt{3}}{2}$.

$$w = \frac{z+i}{iz+2} \quad \text{and} \quad z = x + iy$$

$$\Rightarrow \quad w = \frac{x + iy + i}{ix - y + 2}$$

If z lies on the real axis, then $y = 0$.

So:

$$w = \frac{x+i}{ix+2} = \frac{(x+i)(2-ix)}{4+x^2}$$

$$= \frac{2x - ix^2 + 2i + x}{4+x^2}$$

But

$$w = u + iv$$

So:

$$u = \frac{3x}{4+x^2} \quad \text{and} \quad v = \frac{2-x^2}{4+x^2}$$

If you substitute these values into the left hand side of the given equation of the circle you get:

$$2u^2 + 2v^2 + v - 1 = \frac{18x^2}{(4+x^2)^2} + \frac{8 - 8x^2 + 2x^4}{(4+x^2)^2} + \frac{2 - x^2}{4 + x^2} - 1$$

$$= \frac{18x^2 + 8 - 8x^2 + 2x^4 + 8 - 4x^2 + 2x^2 - x^4 - 16 - 8x^2 - x^4}{(4+x^2)^2}$$

$$= 0$$

So w lies on the circle with equation $2u^2 + 2v^2 + v - 1 = 0$.

Exercise 2D

1 Shade on an Argand diagram the region represented by:

(a) $|z - 3| < 2$ (b) $|z + 2i| > 4$

(c) $|z + 3 - 2i| < 5$ (d) $|z - 5 - i| > 2$

(e) $2 < |z - 3 - 2i| < 3$ (f) $1 < |z + 4 - i| < 5$

(g) $|z - i| < |z - 1|$ (h) $|z - 2i| < |z + 3 - i|$

(i) $|z| < 5|z - 4|$ (j) $|z| > 5|z + 6|$

2 If z_1 and z_2 are complex numbers, show geometrically that

$$|z_1 - z_2| \leqslant |z_1| + |z_2|$$

3 If z_1 and z_2 are complex numbers, show geometrically that

$$|z_1 - z_2| \geqslant ||z_1| - |z_2||$$

4 For the transformation $w = z^2$, show that

(a) as z moves once round the circle centre O and radius 2, w moves twice round the circle centre O and radius 4

(b) as z moves along the imaginary axis, w moves along the negative real axis

(c) as z moves along the real axis, w moves along the positive real axis.

5 For the transformation

$$w = \frac{z + i}{z - i}, \quad z \neq i$$

show that as z moves along the real axis, w moves along a circle centre O and radius 1.

6 Given that $z = \cos\theta + \mathrm{i}\sin\theta$, show that

$$\frac{1}{1-z} = \frac{1}{2} + \frac{\mathrm{i}}{2}\cot\tfrac{1}{2}\theta$$

The point z describes a circle of unit radius with centre O.
Show that for the transformation

$$w = \frac{1}{1-z}, \quad z \neq 1$$

the point w describes a straight line as z describes the circle.

7 For the transformation

$$w = \frac{225}{z}, z \neq 0$$

show that as z moves on the locus $|z - 25| = 25$, w lies on the
locus $|w - 9| = |w|$ and identify this locus geometrically.

8 For the transformation

$$w = \frac{2+z}{\mathrm{i}-z}, z \neq \mathrm{i}$$

find the locus of w as z moves on the circle $|z| = 1$.

9 For the transformation $w = 2z + 3 + \mathrm{i}$, find the locus of w as z
moves on the circle $|z| = 4$.

10 Given that the point z moves around the circle $|z| = 1$ in an
anticlockwise direction starting at $(1, 0)$, describe the path of
the point represented by w where $w =$

(a) $2z^2$ (b) $\frac{1}{z}, z \neq 0$ (c) $\frac{1}{z-2}, z \neq 2$

11 Show that the transformation given by

$$\frac{w+\mathrm{i}}{w-\mathrm{i}} = \frac{z+1}{z-1}, w \neq \mathrm{i}, z \neq 1$$

transforms the real axis of the z-plane into the imaginary axis
of the w-plane.

12 For the transformation

$$w = \frac{z}{z-1}, z \neq 1$$

show that the straight line with equation $x = \tfrac{1}{2}$ in the z-plane
transforms into a circle in the w-plane and find the equation of
the circle.

13 Shade on an Argand diagram the region R for which $|z| < 1$. If z is any point in the region R, and z^* is the conjugate of z, find the corresponding regions for w where:
(a) $w = z + 3 + 4i$ (b) $|wz| = 1$ (c) $w = zz^*$ [E]

14 Points P and Q represent the complex numbers w and z respectively in the Argand diagram. If $w = \dfrac{1 + zi}{z + i}$, $z \neq -i$ and $w = u + iv$, $z = x + iy$, express u and v in terms of x and y. Prove that when P describes the portion of the imaginary axis between the points representing $-i$ and i, Q describes the whole of the positive imaginary axis.   [E]

15 The transformation

$$w = \frac{z + 2}{z + i}$$

where $z \neq -i$, $w \neq 1$, maps the complex number $z = x + iy$ onto the complex number $w = u + iv$.
(a) Show that, if the point representing w lies on the real axis, the point representing z lies on a straight line.
(b) Show further that, if the point representing w lies on the imaginary axis then the point representing z lies on the circle

$$\left| z + 1 + \tfrac{1}{2}i \right| = \tfrac{1}{2}\sqrt{5}$$ [E]

16 Shade and label on an Argand diagram the region R for which

$$|z| \leqslant |z - i|$$

Given that

$$w = \frac{2z - 3i}{2}$$

shade and label on another Argand diagram the region S of the z-plane for which

$$|w| \leqslant |w - i|$$ [E]

17 The transformation $T : z \longmapsto w$ in the complex plane is defined by

$$w = \frac{az + b}{z + c}$$

where $a, b, c \in \mathbb{R}$.
Given that $w = 3$ when $z = 0$ and that $w = 2 - i$ when $z = -1 + i$, find the values of the constants a, b and c.   [E]

18 The transformation

$$w = \frac{z + 2i}{iz + 2}$$

where $z \neq 2i$, maps the complex number $z = x + iy$ onto the complex number $w = u + iv$. Find the complex numbers representing the two points in the complex plane which are invariant under the transformation, giving your answers in the form $a + ib$, where $a, b \in \mathbb{R}$.

Show that, if the point representing z lies on the imaginary axis, then the point representing w also lies on the imaginary axis.

Show further that, if the point representing z lies on the real axis, then the point representing w lies on the circle

$$u^2 + v^2 = 1$$

Sketch this circle, and indicate clearly onto which part of the w-plane the positive real axis of the z-plane is mapped.　　　[E]

SUMMARY OF KEY POINTS

1 The complex number $z = a + ib$, $a, b \in \mathbb{R}$ can also be written as

$$z = r(\cos\theta + i\sin\theta) \quad \text{and} \quad z = re^{i\theta}$$

where $-\pi < \theta \leqslant \pi$ and where $r = \sqrt{(a^2 + b^2)}$ and θ is the angle which the line representing z on an Argand diagram makes with the positive x-axis.

2 $\cos iz = \cosh z$
$\sin iz = i\sinh z$
$\cosh iz = \cos z$
$\sinh iz = i\sin z$
where $z \in \mathbb{C}$.

3 If $z_1 = r_1(\cos\theta_1 + i\sin\theta_1)$ and $z_2 = r_2(\cos\theta_2 + i\sin\theta_2)$ then

$$z_1 z_2 = r_1 r_2 [\cos(\theta_1 + \theta_2) + i\sin(\theta_1 + \theta_2)]$$

$$\frac{z_1}{z_2} = \frac{r_1}{r_2} [\cos(\theta_1 - \theta_2) + i\sin(\theta_1 - \theta_2)]$$

4 De Moivre's theorem states that if
$$z = r(\cos\theta + i\sin\theta)$$

then
$$z^n = r^n(\cos n\theta + i\sin n\theta)$$

5 The nth roots of unity are such that
  (i) one root is always 1
  (ii) if n is even, another root is -1
  (iii) there are n roots and if these are represented on an Argand diagram then the angle between any two consecutive roots is $\dfrac{2\pi}{n}$
  (iv) the roots can be written as $1, \omega, \omega^2, \omega^3, \ldots, \omega^{n-2}, \omega^{n-1}$
  (v) with the exception of 1 (and -1 if n is even) the roots occur in conjugate pairs
  (vi) $1 + \omega + \omega^2 + \omega^3 + \ldots + \omega^{n-2} + \omega^{n-1} = 0$

6 $|z_1 + z_2| \leqslant |z_1| + |z_2|$
  $|z_1 + z_2| \geqslant \big||z_1| - |z_2|\big|$

Matrix algebra

3

So far during this course you should have learned quite a lot of algebra. In this chapter you will extend that knowledge. You will be introduced to the idea of a linear transformation and its inverse and learn how to represent such a transformation by a matrix. You will then learn how to find the inverse of a matrix so that you can represent an inverse transformation. Finally, the chapter will introduce the idea of eigenvectors and eigenvalues and you will find out how to diagonalise a symmetric matrix using eigenvalues and eigenvectors.

3.1 Linear transformations

Chapter 2 of Book C3 introduced mappings and functions. It described mappings such as $f : x \mapsto 3x^2 + 2$. So a mapping is a rule which, if you apply it to a number, gives you another number. For example,

$$f(2) = 3 \times 2^2 + 2 = 14$$
$$f(-6) = 3(-6)^2 + 2 = 110$$

The sets of numbers which you can possibly feed in to the mapping is called the **domain** of the mapping.

A **transformation** is also a rule which has a domain. However, unlike mappings and functions, the domain of a transformation is a set of vectors. So, whereas the domain of a function may be $\mathbb{R}$, the set of real numbers, the domain of a transformation in this chapter is always going to be a set of *vectors* in either two or three dimensions.

Vectors involved with transformations are written in column vector form as (2×1) or (3×1) matrices; for example:

$$x\mathbf{i} + y\mathbf{j} \equiv \begin{pmatrix} x \\ y \end{pmatrix} \quad \text{and} \quad x\mathbf{i} + y\mathbf{j} + z\mathbf{k} \equiv \begin{pmatrix} x \\ y \\ z \end{pmatrix}$$

Notice that in two dimensions

$$\mathbf{i} \equiv \begin{pmatrix} 1 \\ 0 \end{pmatrix} \qquad \mathbf{j} \equiv \begin{pmatrix} 0 \\ 1 \end{pmatrix}$$

and in three dimensions

$$\mathbf{i} \equiv \begin{pmatrix} 1 \\ 0 \\ 0 \end{pmatrix} \qquad \mathbf{j} \equiv \begin{pmatrix} 0 \\ 1 \\ 0 \end{pmatrix} \qquad \mathbf{k} \equiv \begin{pmatrix} 0 \\ 0 \\ 1 \end{pmatrix}$$

So a transformation T may be defined as

$$T : \mathbf{v} \mapsto -\mathbf{v}$$

That is, as an example in two dimensions,

$$T\begin{pmatrix} x \\ y \end{pmatrix} = \begin{pmatrix} -x \\ -y \end{pmatrix}, \quad \text{where } \begin{pmatrix} x \\ y \end{pmatrix} \equiv x\mathbf{i} + y\mathbf{j}$$

As another example, in three dimensions,

$$T : \begin{pmatrix} x \\ y \\ z \end{pmatrix} \mapsto \begin{pmatrix} x + y \\ y + z \\ z + x \end{pmatrix}, \quad \text{where } \begin{pmatrix} x \\ y \\ z \end{pmatrix} \equiv x\mathbf{i} + y\mathbf{j} + z\mathbf{k}$$

A transformation is called *linear* if it is such that

$$T(\mathbf{v}_1 + \mathbf{v}_2) = T(\mathbf{v}_1) + T(\mathbf{v}_2)$$

and

$$T(a\mathbf{v}) = aT(\mathbf{v})$$

where a is a scalar.

These two conditions can be put together and you then get a single condition for a transformation T to be linear, which is:

■ $$T(a_1\mathbf{v}_1 + a_2\mathbf{v}_2) = a_1 T(\mathbf{v}_1) + a_2 T(\mathbf{v}_2)$$

where a_1, a_2 are scalars.

In general, for a transformation T, where $T : \mathbf{v} \mapsto T(\mathbf{v})$, the set of vectors V_1 being transformed, of which $\mathbf{v}_1$ and $\mathbf{v}_2$ are typical members, is called the **domain** of T. The set of vectors V_2 obtained under the transformation, of which $T(\mathbf{v}_1)$ and $T(\mathbf{v}_2)$ are the members corresponding to $\mathbf{v}_1$ and $\mathbf{v}_2$ respectively, is called the **image** of T. You can represent the conditions described above diagramatically like this:

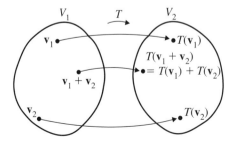

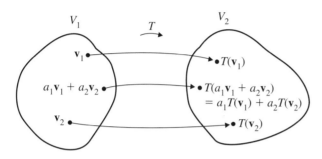

You can represent the two sets of conditions combined like this:

Example 1

Show that T, where

$$T : \begin{pmatrix} x \\ y \end{pmatrix} \mapsto \begin{pmatrix} x \\ -y \end{pmatrix}$$

is a linear transformation.

If $\mathbf{v}_1 = \begin{pmatrix} x_1 \\ y_1 \end{pmatrix}$ and $\mathbf{v}_2 = \begin{pmatrix} x_2 \\ y_2 \end{pmatrix}$ then

$$a_1 \mathbf{v}_1 + a_2 \mathbf{v}_2 = \begin{pmatrix} a_1 x_1 \\ a_1 y_1 \end{pmatrix} + \begin{pmatrix} a_2 x_2 \\ a_2 y_2 \end{pmatrix}$$

So:

$$T(a_1 \mathbf{v}_1 + a_2 \mathbf{v}_2) = T\left[\begin{pmatrix} a_1 x_1 \\ a_1 y_1 \end{pmatrix} + \begin{pmatrix} a_2 x_2 \\ a_2 y_2 \end{pmatrix} \right]$$

$$= T \begin{pmatrix} a_1 x_1 + a_2 x_2 \\ a_1 y_1 + a_2 y_2 \end{pmatrix}$$

$$= \begin{pmatrix} a_1 x_1 + a_2 x_2 \\ -a_1 y_1 - a_2 y_2 \end{pmatrix}$$

$$= \begin{pmatrix} a_1 x_1 \\ -a_1 y_1 \end{pmatrix} + \begin{pmatrix} a_2 x_2 \\ -a_2 y_2 \end{pmatrix}$$

$$= a_1 \begin{pmatrix} x_1 \\ -y_1 \end{pmatrix} + a_2 \begin{pmatrix} x_2 \\ -y_2 \end{pmatrix}$$

$$= a_1 T(\mathbf{v}_1) + a_2 T(\mathbf{v}_2)$$

So T is a linear transformation.

Example 2

Show that T is a linear transformation where

$$T: \begin{pmatrix} x \\ y \\ z \end{pmatrix} \mapsto 6\begin{pmatrix} x \\ x - y \\ x + z \end{pmatrix}$$

Let $\mathbf{v}_1 = \begin{pmatrix} x_1 \\ y_1 \\ z_1 \end{pmatrix}$ and $\mathbf{v}_2 = \begin{pmatrix} x_2 \\ y_2 \\ z_2 \end{pmatrix}$

then:
$$a_1\mathbf{v}_1 + a_2\mathbf{v}_2 = \begin{pmatrix} a_1 x_1 \\ a_1 y_1 \\ a_1 z_1 \end{pmatrix} + \begin{pmatrix} a_2 x_2 \\ a_2 y_2 \\ a_2 z_2 \end{pmatrix}$$

$$= \begin{pmatrix} a_1 x_1 + a_2 x_2 \\ a_1 y_1 + a_2 y_2 \\ a_1 z_1 + a_2 z_2 \end{pmatrix}$$

So:
$$T(a_1\mathbf{v}_1 + a_2\mathbf{v}_2) = T\begin{pmatrix} a_1 x_1 + a_2 x_2 \\ a_1 y_1 + a_2 y_2 \\ a_1 z_1 + a_2 z_2 \end{pmatrix}$$

$$= 6\begin{pmatrix} a_1 x_1 + a_2 x_2 \\ a_1 x_1 + a_2 x_2 - a_1 y_1 - a_2 y_2 \\ a_1 x_1 + a_2 x_2 + a_1 z_1 + a_2 z_2 \end{pmatrix}$$

$$= 6\left[\begin{pmatrix} a_1 x_1 \\ a_1 x_1 - a_1 y_1 \\ a_1 x_1 + a_1 z_1 \end{pmatrix} + \begin{pmatrix} a_2 x_2 \\ a_2 x_2 - a_2 y_2 \\ a_2 x_2 + a_2 z_2 \end{pmatrix} \right]$$

$$= 6\left[a_1\begin{pmatrix} x_1 \\ x_1 - y_1 \\ x_1 + z_1 \end{pmatrix} + a_2\begin{pmatrix} x_2 \\ x_2 - y_2 \\ x_2 + z_2 \end{pmatrix} \right]$$

$$= a_1\left[6\begin{pmatrix} x_1 \\ x_1 - y_1 \\ x_1 + z_1 \end{pmatrix} \right] + a_2\left[6\begin{pmatrix} x_2 \\ x_2 - y_2 \\ x_2 + z_2 \end{pmatrix} \right]$$

$$= a_1 T(\mathbf{v}_1) + a_2 T(\mathbf{v}_2)$$

So T is a linear transformation.

3.2 Composite linear transformations

Many functions are composite functions, for example:

$$f : x \mapsto 2x$$

and
$$g : x \mapsto x^2$$

so:
$$fg : x \mapsto 2x^2$$

This also happens with transformations.

If $S : \mathbf{v} \mapsto S(\mathbf{v})$, where $\mathbf{v} \in V_1$ and $S(\mathbf{v}) \in V_2$
and if $T : S(\mathbf{v}) \mapsto T[S(\mathbf{v})]$, where $T[S(\mathbf{v})] \in V_3$
then $TS : \mathbf{v} \mapsto TS(\mathbf{v})$, where T and S are both linear transformations.

That is, diagramatically

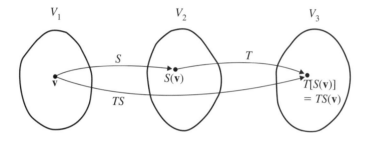

TS is still a linear transformation because

$$TS(a_1\mathbf{v}_1 + a_2\mathbf{v}_2) = T[S(a_1\mathbf{v}_1 + a_2\mathbf{v}_2)]$$
$$= T[a_1 S(\mathbf{v}_1) + a_2 S(\mathbf{v}_2)] \text{ since } S \text{ is linear}$$
$$= a_1 T[S(\mathbf{v}_1)] + a_2 T[S(\mathbf{v}_2)] \text{ since } T \text{ is linear}$$
$$= a_1 TS(\mathbf{v}_1) + a_2 TS(\mathbf{v}_2)$$

As with functions, it is important to notice the order in which the transformations are applied since, in general, $TS \neq ST$. Remember that TS means 'do S first, then T'.

The transformation I, where

$$I : \mathbf{v} \mapsto \mathbf{v}$$

is called the **identity transformation** because it maps each vector to itself. Consequently $I(\mathbf{v})$ is identical to $\mathbf{v}$. You should easily be able to see that

$$IT = TI = T$$

where T is a linear transformation.

3.3 Inverse transformations

Again, as with functions, it is sometimes the case that the transformation is one–one. That is, T maps each $\mathbf{v}$ to only a single vector $T(\mathbf{v})$ in the image. Likewise, there is another transformation T^{-1} which maps each $T(\mathbf{v})$ in the image to a single vector $\mathbf{v}$ in the original domain. That is:

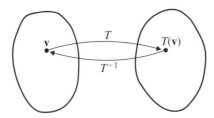

T^{-1} is called the **inverse transformation** of T.

Since $$T : \mathbf{v} \mapsto T(\mathbf{v})$$

and $$T^{-1} : T(\mathbf{v}) \mapsto \mathbf{v}$$

then $$T^{-1}T : \mathbf{v} \mapsto \mathbf{v}$$

So: $$T^{-1}T = I$$

Likewise: $$T^{-1} : T(\mathbf{v}) \mapsto \mathbf{v} \quad \text{and} \quad T : \mathbf{v} \mapsto T(\mathbf{v})$$

so $$TT^{-1} : T(\mathbf{v}) \mapsto T(\mathbf{v})$$

So: $$TT^{-1} = I$$

■ **If T is one–one, then an inverse linear transformation exists which is such that**

$$TT^{-1} = T^{-1}T = I$$

Example 3

Find the inverse transformation T^{-1} of $\ T : \begin{pmatrix} x \\ y \\ z \end{pmatrix} \mapsto \begin{pmatrix} x \\ x+y \\ z \end{pmatrix}$

Let $\begin{pmatrix} p_1 \\ p_2 \\ p_3 \end{pmatrix} = \begin{pmatrix} x \\ x+y \\ z \end{pmatrix}$

then: $\quad x = p_1, z = p_3 \quad \text{and} \quad y = p_2 - x = p_2 - p_1$

So $$T^{-1} : \begin{pmatrix} p_1 \\ p_2 \\ p_3 \end{pmatrix} \mapsto \begin{pmatrix} p_1 \\ p_2 - p_1 \\ p_3 \end{pmatrix}$$

or $$T^{-1} : \begin{pmatrix} x \\ y \\ z \end{pmatrix} \mapsto \begin{pmatrix} x \\ y - x \\ z \end{pmatrix}$$

Exercise 3A

1 Show that the following are linear transformations:

(a) $T: \begin{pmatrix} x \\ y \end{pmatrix} \mapsto \begin{pmatrix} x \\ 2y \end{pmatrix}$

(b) $T: \begin{pmatrix} x \\ y \end{pmatrix} \mapsto 3 \begin{pmatrix} x \\ \frac{1}{2}y \end{pmatrix}$

(c) $T: \begin{pmatrix} x \\ y \end{pmatrix} \mapsto \begin{pmatrix} y \\ x \end{pmatrix}$

(d) $T: \begin{pmatrix} x \\ y \end{pmatrix} \mapsto \begin{pmatrix} 3x \\ -2y \end{pmatrix}$

(e) $T: \begin{pmatrix} x \\ y \\ z \end{pmatrix} \mapsto \begin{pmatrix} 2x \\ 3y \\ -z \end{pmatrix}$

(f) $T: \begin{pmatrix} x \\ y \\ z \end{pmatrix} \mapsto \begin{pmatrix} x + 2y \\ y + z \\ x - 2y \end{pmatrix}$

(g) $T: \begin{pmatrix} x \\ y \\ z \end{pmatrix} \mapsto \begin{pmatrix} x - y \\ z \\ 0 \end{pmatrix}$

(h) $T: \begin{pmatrix} x \\ y \\ z \end{pmatrix} \mapsto \begin{pmatrix} 3x - 2y \\ 2y - z \end{pmatrix}$

(i) $T: \begin{pmatrix} x \\ y \\ z \end{pmatrix} \mapsto \begin{pmatrix} x + y + z \\ x + y + z \\ x + y + z \end{pmatrix}$

(j) $T: \begin{pmatrix} x \\ y \\ z \end{pmatrix} \mapsto \begin{pmatrix} 2x - 2y + z \\ 3x - y + 2z \\ 6x - y - z \end{pmatrix}$

2 Find T^{-1}, where possible, in the following cases:

(a) $T: \begin{pmatrix} x \\ y \end{pmatrix} \mapsto \begin{pmatrix} x \\ 2y \end{pmatrix}$

(b) $T: \begin{pmatrix} x \\ y \end{pmatrix} \mapsto \begin{pmatrix} y \\ x \end{pmatrix}$

(c) $T: \begin{pmatrix} x \\ y \end{pmatrix} \mapsto \begin{pmatrix} x \\ -y \end{pmatrix}$

(d) $T: \begin{pmatrix} x \\ y \end{pmatrix} \mapsto \begin{pmatrix} x \\ 0 \end{pmatrix}$

(e) $T: \begin{pmatrix} x \\ y \end{pmatrix} \mapsto \begin{pmatrix} x + 2y \\ y + 3x \end{pmatrix}$

(f) $T: \begin{pmatrix} x \\ y \end{pmatrix} \mapsto \begin{pmatrix} x + 2y \\ x + 2y \end{pmatrix}$

(g) $T: \begin{pmatrix} x \\ y \\ z \end{pmatrix} \mapsto \begin{pmatrix} x \\ x + y \\ y + z \end{pmatrix}$

(h) $T: \begin{pmatrix} x \\ y \\ z \end{pmatrix} \mapsto \begin{pmatrix} x \\ 2y \\ 0 \end{pmatrix}$

(i) $T: \begin{pmatrix} x \\ y \\ z \end{pmatrix} \mapsto \begin{pmatrix} x + y \\ x + y + z \\ 2y - z \end{pmatrix}$

(j) $T: \begin{pmatrix} x \\ y \\ z \end{pmatrix} \mapsto \begin{pmatrix} 2x - y \\ x + z \end{pmatrix}$

3.4 Matrices

A **matrix** is an array of numbers such as

$$\begin{pmatrix} 1 & 2 \\ 0 & -1 \end{pmatrix}, \begin{pmatrix} 1 & 2 & -1 \\ 0 & 1 & 2 \\ 1 & 1 & 0 \end{pmatrix} \text{ and } \begin{pmatrix} 1 & 2 & 1 \\ 0 & -1 & 2 \end{pmatrix}$$

Because the first matrix has 2 rows and 2 columns we call it a 2×2 ('2 by 2') matrix. The second matrix has 3 rows and 3 columns so you say it is a 3×3 matrix and the third matrix is called a 2×3 matrix because it has 2 rows and 3 columns. The description 2×2, 3×3, 2×3 and so on is called the **order** of the matrix.

Adding and subtracting matrices

You can add or subtract two matrices so long as they have the same order. So you can add a 2×3 matrix to a 2×3 matrix and you can add a 3×3 matrix to a 3×3 matrix, and so on.

Each of the numbers in a matrix is called an **element** and you add two matrices by adding their corresponding elements. For example:

$$\begin{pmatrix} 2 & 1 \\ 3 & -4 \end{pmatrix} + \begin{pmatrix} 6 & -5 \\ 1 & -7 \end{pmatrix} = \begin{pmatrix} 2+6 & 1-5 \\ 3+1 & -4-7 \end{pmatrix}$$
$$= \begin{pmatrix} 8 & -4 \\ 4 & -11 \end{pmatrix}$$

You subtract one matrix from another by subtracting the corresponding elements. For example:

$$\begin{pmatrix} 1 & 0 & -1 \\ 2 & 1 & 2 \\ 3 & 1 & -2 \end{pmatrix} - \begin{pmatrix} 0 & 1 & 1 \\ 2 & 1 & -3 \\ 0 & 1 & 2 \end{pmatrix} = \begin{pmatrix} 1-0 & 0-1 & -1-1 \\ 2-2 & 1-1 & 2-(-3) \\ 3-0 & 1-1 & -2-2 \end{pmatrix}$$
$$= \begin{pmatrix} 1 & -1 & -2 \\ 0 & 0 & 5 \\ 3 & 0 & -4 \end{pmatrix}$$

Multiplying matrices

You multiply two matrices like this:

$$\begin{pmatrix} 1 & 0 & -1 \\ 2 & 1 & 2 \\ 3 & 1 & -2 \end{pmatrix} \begin{pmatrix} 0 & 1 & 1 \\ 2 & 1 & 3 \\ 0 & 1 & 2 \end{pmatrix}$$

Turn the first row of the first matrix to make it a vertical column and then place it beside the first column in the second matrix. Then multiply the corresponding elements and add the answers.

So:
$$
\begin{bmatrix} 0 \\ 2 \\ 0 \end{bmatrix} \times \begin{bmatrix} 1 \\ 0 \\ -1 \end{bmatrix} = \begin{matrix} 0 \\ + \\ 0 \\ + \\ 0 \end{matrix} = 0
$$

Because you obtained this answer by multiplying the *first row* of the first matrix by the *first column* of the second matrix, the answer goes in the *first row* and *first column* of the answer matrix. So the answer matrix starts like this:

$$
\begin{pmatrix} 0 & \\ & \end{pmatrix}
$$

Now turn the first row of the first matrix into a column and place it against the *second* column of the second matrix. So you get:

$$
\begin{bmatrix} 1 \\ 1 \\ 1 \end{bmatrix} \times \begin{bmatrix} 1 \\ 0 \\ -1 \end{bmatrix} = \begin{matrix} 1 \\ + \\ 0 \\ + \\ -1 \end{matrix} = 0
$$

The answer goes in the *first row* and *second column* of the answer matrix because it comes from multiplying the first row of the first matrix by the second column of the second matrix. The answer matrix now looks like this:

$$
\begin{pmatrix} 0 & 0 \\ & \end{pmatrix}
$$

Now move the first row against the *third* column and get:

$$
\begin{bmatrix} 1 \\ 3 \\ 2 \end{bmatrix} \times \begin{bmatrix} 1 \\ 0 \\ -1 \end{bmatrix} = \begin{matrix} 1 \\ + \\ 0 \\ + \\ -2 \end{matrix} = -1
$$

The answer matrix now looks like this:

$$\begin{pmatrix} 0 & 0 & -1 \\ & & \\ & & \end{pmatrix}$$

$$\begin{pmatrix} 1 & 0 & -1 \\ 2 & 1 & 2 \\ 3 & 1 & -2 \end{pmatrix} \begin{pmatrix} 0 & 1 & 1 \\ 2 & 1 & 3 \\ 0 & 1 & 2 \end{pmatrix}$$

Now take the *second* row of the first matrix and repeat the process.

$$\begin{matrix} 0 & & 2 & & 0 \\ & & & & + \\ 2 & \times & 1 & = & 2 & = 2 \\ & & & & + \\ 0 & & 2 & & 0 \end{matrix}$$

$$\begin{matrix} 1 & & 2 & & 2 \\ & & & & + \\ 1 & \times & 1 & = & 1 & = 5 \\ & & & & + \\ 1 & & 2 & & 2 \end{matrix}$$

$$\begin{matrix} 1 & & 2 & & 2 \\ & & & & + \\ 3 & \times & 1 & = & 3 & = 9 \\ & & & & + \\ 2 & & 2 & & 4 \end{matrix}$$

The answer matrix now looks like this:

$$\begin{pmatrix} 0 & 0 & -1 \\ 2 & 5 & 9 \\ & & \end{pmatrix}$$

Finally, repeat the process with the *third* row of the first matrix.

$$\begin{matrix} 0 & & 3 & & 0 \\ & & & & + \\ 2 & \times & 1 & = & 2 & = 2 \\ & & & & + \\ 0 & & -2 & & 0 \end{matrix}$$

$$\begin{matrix} 1 & & 3 & & 3 \\ & & & & + \\ 1 & \times & 1 & = & 1 & = 2 \\ & & & & + \\ 1 & & -2 & & -2 \end{matrix}$$

$$\begin{matrix} 1 \\ 3 \\ 2 \end{matrix} \times \begin{matrix} 3 \\ 1 \\ -2 \end{matrix} = \begin{matrix} 3 \\ + \\ 3 \\ + \\ -4 \end{matrix} = 2$$

The answer is therefore

$$\begin{pmatrix} 0 & 0 & -1 \\ 2 & 5 & 9 \\ 2 & 2 & 2 \end{pmatrix}$$

Because when you multiply two matrices such as

$$\begin{pmatrix} 1 & 0 & -1 \\ 2 & 1 & 2 \\ 3 & 1 & -2 \end{pmatrix} \begin{pmatrix} 0 & 1 & 1 \\ 2 & 1 & 3 \\ 0 & 1 & 2 \end{pmatrix}$$

you first of all place the $\begin{matrix} 1 & 0 & -1 \end{matrix}$ against the

$\begin{matrix} 0 \\ 2 \\ 0 \end{matrix}$, you can see that the first matrix *must* have the same number of

columns as the second matrix has rows.

If you reverse the order of the two matrices above and multiply, you get:

$$\begin{pmatrix} 0 & 1 & 1 \\ 2 & 1 & 3 \\ 0 & 1 & 2 \end{pmatrix} \begin{pmatrix} 1 & 0 & -1 \\ 2 & 1 & 2 \\ 3 & 1 & -2 \end{pmatrix} = \begin{pmatrix} 5 & 2 & 0 \\ 13 & 4 & -6 \\ 8 & 3 & -2 \end{pmatrix}$$

This, of course, is different to the answer you obtained before. **It is *very* important therefore that you remember that if you multiply two matrices in reverse order, in general you do not get the same answer.** Matrix multiplication is *not* commutative.

Square matrices

Any matrix which has the same number of rows as columns is called a **square matrix**.

So $\begin{pmatrix} 1 & 3 \\ 2 & -3 \end{pmatrix}$, $\begin{pmatrix} 2 & -1 & 0 \\ 1 & 4 & 6 \\ -1 & 2 & 5 \end{pmatrix}$, $\begin{pmatrix} 2 & 1 & 3 & 5 & 4 \\ -1 & 2 & 0 & 6 & 1 \\ 1 & 2 & 7 & 3 & 9 \\ -1 & 3 & 1 & 5 & 8 \\ 8 & 3 & 2 & 1 & 5 \end{pmatrix}$

are all square matrices of order 2×2, 3×3 and 5×5 respectively.

The identity matrix

If you have the matrix $\begin{pmatrix} 2 & 4 \\ -3 & 2 \end{pmatrix}$ and multiply it by $\begin{pmatrix} 1 & 0 \\ 0 & 1 \end{pmatrix}$ you get:

$$\begin{pmatrix} 2 & 4 \\ -3 & 2 \end{pmatrix} \begin{pmatrix} 1 & 0 \\ 0 & 1 \end{pmatrix} = \begin{pmatrix} 2 & 4 \\ -3 & 2 \end{pmatrix}$$

Also:

$$\begin{pmatrix} 1 & 0 \\ 0 & 1 \end{pmatrix} \begin{pmatrix} 2 & 4 \\ -3 & 2 \end{pmatrix} = \begin{pmatrix} 2 & 4 \\ -3 & 2 \end{pmatrix}$$

If you take the matrix $\begin{pmatrix} 1 & 2 & -1 \\ 3 & 1 & -4 \\ -4 & 1 & 2 \end{pmatrix}$ and multiply it by

$\begin{pmatrix} 1 & 0 & 0 \\ 0 & 1 & 0 \\ 0 & 0 & 1 \end{pmatrix}$ you get

$$\begin{pmatrix} 1 & 2 & -1 \\ 3 & 1 & -4 \\ -4 & 1 & 2 \end{pmatrix} \begin{pmatrix} 1 & 0 & 0 \\ 0 & 1 & 0 \\ 0 & 0 & 1 \end{pmatrix} = \begin{pmatrix} 1 & 2 & -1 \\ 3 & 1 & -4 \\ -4 & 1 & 2 \end{pmatrix}$$

Also:

$$\begin{pmatrix} 1 & 0 & 0 \\ 0 & 1 & 0 \\ 0 & 0 & 1 \end{pmatrix} \begin{pmatrix} 1 & 2 & -1 \\ 3 & 1 & -4 \\ -4 & 1 & 2 \end{pmatrix} = \begin{pmatrix} 1 & 2 & -1 \\ 3 & 1 & -4 \\ -4 & 1 & 2 \end{pmatrix}$$

The matrix $\begin{pmatrix} 1 & 0 \\ 0 & 1 \end{pmatrix}$ is called the 2×2 **identity matrix** because when you multiply any 2×2 matrix **A** by $\begin{pmatrix} 1 & 0 \\ 0 & 1 \end{pmatrix}$ you get the answer **A**.

The matrix $\begin{pmatrix} 1 & 0 & 0 \\ 0 & 1 & 0 \\ 0 & 0 & 1 \end{pmatrix}$ is called the 3×3 identity matrix because when you multiply any 3×3 matrix **B** by $\begin{pmatrix} 1 & 0 & 0 \\ 0 & 1 & 0 \\ 0 & 0 & 1 \end{pmatrix}$ you get the answer **B**.

The identity matrix is usually called **I**, whether it is the 2×2 identity matrix, the 3×3 identity matrix or whatever. It should be obvious from the context which identity matrix **I** refers to.

■ **You should remember that**
$$\mathbf{AI} = \mathbf{IA} = \mathbf{A}$$
In the case of the identity matrix, multiplication *is* commutative.

Identity matrices only exist for *square* matrices.

3.5 The inverse of a matrix

If you have two square matrices **A** and **B** such that

$$\mathbf{AB} = \mathbf{I}$$

then **A** is called the **inverse** of **B** and **B** is called the **inverse** of **A** so:

$$\begin{pmatrix} 3 & 2 \\ 4 & 3 \end{pmatrix}\begin{pmatrix} 3 & -2 \\ -4 & 3 \end{pmatrix} = \begin{pmatrix} 1 & 0 \\ 0 & 1 \end{pmatrix}$$

Therefore $\begin{pmatrix} 3 & 2 \\ 4 & 3 \end{pmatrix}$ is the inverse of $\begin{pmatrix} 3 & -2 \\ -4 & 3 \end{pmatrix}$ and $\begin{pmatrix} 3 & -2 \\ -4 & 3 \end{pmatrix}$ is

the inverse of $\begin{pmatrix} 3 & 2 \\ 4 & 3 \end{pmatrix}$.

If **A** is the inverse of **B** and vice versa you write

$$\mathbf{A} = \mathbf{B}^{-1} \quad \text{and} \quad \mathbf{B} = \mathbf{A}^{-1}$$

You should notice that in the definition $\mathbf{AB} = \mathbf{I}$, **A** and **B** are both square. It is only square matrices that can have an inverse.

Inverse of a 2 × 2 matrix

To find the inverse of a 2×2 matrix is quite easy. If $\mathbf{A} = \begin{pmatrix} p & q \\ r & s \end{pmatrix}$

then first of all swap the p and the s, then change the signs of the q and the r.

Thus you get: $\begin{pmatrix} s & -q \\ -r & p \end{pmatrix}$

Now divide every element in this matrix by the determinant of **A**.

■ **The determinant of A, usually written as det A, is $ps - qr$.**

So: $\quad \mathbf{A}^{-1} = \begin{pmatrix} \dfrac{s}{ps-qr} & \dfrac{-q}{ps-qr} \\ \dfrac{-r}{ps-qr} & \dfrac{p}{ps-qr} \end{pmatrix} = \dfrac{1}{ps-qr}\begin{pmatrix} s & -q \\ -r & p \end{pmatrix}$

■ **That is, the inverse of a 2 × 2 matrix $\begin{pmatrix} p & q \\ r & s \end{pmatrix}$ is:**

$$\begin{pmatrix} p & q \\ r & s \end{pmatrix}^{-1} = \frac{1}{ps-qr}\begin{pmatrix} s & -q \\ -r & p \end{pmatrix}$$

If $ps - qr = 0$, the matrix is called **singular** and there is *no* inverse

matrix of $\begin{pmatrix} p & q \\ r & s \end{pmatrix}$ in this special case.

Example 4

Find the inverse of

$$\mathbf{A} = \begin{pmatrix} 2 & -3 \\ -2 & 4 \end{pmatrix}$$

det $\mathbf{A} = (2 \times 4) - (-3 \times -2) = 8 - 6 = 2$

So $\mathbf{A}^{-1} = \frac{1}{2} \begin{pmatrix} 4 & 3 \\ 2 & 2 \end{pmatrix}$

As a check, calculate $\mathbf{A}^{-1}\mathbf{A}$:

$$\frac{1}{2} \begin{pmatrix} 4 & 3 \\ 2 & 2 \end{pmatrix} \begin{pmatrix} 2 & -3 \\ -2 & 4 \end{pmatrix} = \frac{1}{2} \begin{pmatrix} 2 & 0 \\ 0 & 2 \end{pmatrix}$$

$$= \begin{pmatrix} 1 & 0 \\ 0 & 1 \end{pmatrix} = \mathbf{I}$$

Example 5

Find the inverse of

$$\mathbf{B} = \begin{pmatrix} -2 & 1 \\ -2 & 3 \end{pmatrix}$$

det $\mathbf{B} = (-2 \times 3) - (1 \times -2) = -6 + 2 = -4$

So $\mathbf{B}^{-1} = -\frac{1}{4} \begin{pmatrix} 3 & -1 \\ 2 & -2 \end{pmatrix}$

As a check, calculate $\mathbf{B}^{-1}\mathbf{B}$:

$$-\frac{1}{4} \begin{pmatrix} 3 & -1 \\ 2 & -2 \end{pmatrix} \begin{pmatrix} -2 & 1 \\ -2 & 3 \end{pmatrix} = -\frac{1}{4} \begin{pmatrix} -4 & 0 \\ 0 & -4 \end{pmatrix}$$

$$= \begin{pmatrix} 1 & 0 \\ 0 & 1 \end{pmatrix} = \mathbf{I}$$

Transpose of a matrix

If you have a matrix and take its first row and write this as the first column, then take its second row and write this as the second column, and so on, the resulting matrix is called the **transpose** of the first. If

$$\mathbf{A} = \begin{pmatrix} 2 & 1 & 2 \\ 1 & 4 & 6 \\ 1 & -1 & 2 \end{pmatrix} \text{ then the transpose of } \mathbf{A} \text{ is } \begin{pmatrix} 2 & 1 & 1 \\ 1 & 4 & -1 \\ 2 & 6 & 2 \end{pmatrix}$$

You write the transpose of $\mathbf{A}$ as $\mathbf{A}^{\mathrm{T}}$.

Example 6
Find the transpose of

$$\mathbf{A} = \begin{pmatrix} 1 & -1 & 2 \\ 5 & 7 & 9 \\ -3 & 1 & 4 \end{pmatrix}$$

$$\mathbf{A}^{\mathrm{T}} = \begin{pmatrix} 1 & 5 & -3 \\ -1 & 7 & 1 \\ 2 & 9 & 4 \end{pmatrix}$$

Inverse of a 3 × 3 matrix

If you need to find the inverse of a 3×3 matrix this is much more complicated than finding the inverse of a 2×2 matrix. However, as with the 2×2 matrix, you do need to find the determinant. This, in itself, is much more complicated than in the 2×2 case.

Consider the matrix

$$\mathbf{A} = \begin{pmatrix} 1 & 2 & 4 \\ -1 & 3 & 0 \\ 0 & 1 & 5 \end{pmatrix}$$

To find det $\mathbf{A}$, which you can also write as

$$\begin{vmatrix} 1 & 2 & 4 \\ -1 & 3 & 0 \\ 0 & 1 & 5 \end{vmatrix}$$

you take the number at the top of the first column (in this case it is 1). Then you put a line through the row and through the column in which the 1 lies. That is:

$$\begin{pmatrix} 1 & 2 & 4 \\ -1 & 3 & 0 \\ 0 & 1 & 5 \end{pmatrix}$$

Now you find the determinant of what remains:

$$\begin{vmatrix} 3 & 0 \\ 1 & 5 \end{vmatrix} = (3 \times 5) - (0 \times 1) = 15$$

Now multiply the 1 at the top of the first column of the matrix by this determinant:

$$1 \times \begin{vmatrix} 3 & 0 \\ 1 & 5 \end{vmatrix} = 1 \times 15 = 15$$

At this point you repeat the process with the number at the top of the second column (which in this case is 2). So you cross out the row and column in which the 2 lies:

$$\begin{pmatrix} \cancel{1} & \cancel{2} & \cancel{4} \\ -1 & \cancel{3} & 0 \\ 0 & \cancel{1} & 5 \end{pmatrix}$$

Then you find the determinant of what remains:

$$\begin{vmatrix} -1 & 0 \\ 0 & 5 \end{vmatrix} = (-1 \times 5) - (0 \times 0) = -5$$

Now multiply the 2 at the top of the second column of the matrix by this determinant *and also change the sign*:

$$-2 \times \begin{vmatrix} -1 & 0 \\ 0 & 5 \end{vmatrix} = (-2) \times (-5) = 10$$

Finally, take the number at the top of the third column (in this case 4) and cross out the row and the column in which it lies:

$$\begin{pmatrix} \cancel{1} & \cancel{2} & \cancel{4} \\ -1 & 3 & \cancel{0} \\ 0 & 1 & \cancel{5} \end{pmatrix}$$

The determinant of what is left is

$$\begin{vmatrix} -1 & 3 \\ 0 & 1 \end{vmatrix} = (-1 \times 1) - (3 \times 0) = -1$$

So you multiply the 4 by the determinant to get

$$4 \begin{vmatrix} -1 & 3 \\ 0 & 1 \end{vmatrix} = 4 \times (-1) = -4$$

The determinant of

$$\begin{pmatrix} 1 & 2 & 4 \\ -1 & 3 & 0 \\ 0 & 1 & 5 \end{pmatrix}$$

is then the sum of these three answers. That is:

$$\begin{vmatrix} 1 & 2 & 4 \\ -1 & 3 & 0 \\ 0 & 1 & 5 \end{vmatrix} = 15 + 10 - 4 = 21$$

To find the general formula for the determinant of a 3×3 matrix, consider the matrix

$$\begin{pmatrix} a & b & c \\ d & e & f \\ g & h & i \end{pmatrix}$$

If you take a and cross out the row and column in which it lies you get:

$$\begin{pmatrix} a & b & c \\ d & e & f \\ g & h & i \end{pmatrix}$$

So you first of all calculate

$$a \times \begin{vmatrix} e & f \\ h & i \end{vmatrix}$$

Then take b and cross out the row and column in which it lies:

$$\begin{pmatrix} a & b & c \\ d & e & f \\ g & h & i \end{pmatrix}$$

Now you calculate

$$-b \times \begin{vmatrix} d & f \\ g & i \end{vmatrix}$$

(Remember you *must* change the sign here.)

Finally, take c and cross out the row and column in which it lies:

$$\begin{pmatrix} a & b & c \\ d & e & f \\ g & h & i \end{pmatrix}$$

So the third calculation you have to do is

$$c \times \begin{vmatrix} d & e \\ g & h \end{vmatrix}$$

So:

$$\blacksquare \qquad \begin{vmatrix} a & b & c \\ d & e & f \\ g & h & i \end{vmatrix} = a \begin{vmatrix} e & f \\ h & i \end{vmatrix} - b \begin{vmatrix} d & f \\ g & i \end{vmatrix} + c \begin{vmatrix} d & e \\ g & h \end{vmatrix}$$

Example 7

Find the determinant of

$$\begin{pmatrix} 1 & 2 & 7 \\ 3 & -5 & 2 \\ 1 & 1 & 4 \end{pmatrix}$$

$$\begin{vmatrix} 1 & 2 & 7 \\ 3 & -5 & 2 \\ 1 & 1 & 4 \end{vmatrix} = 1 \begin{vmatrix} -5 & 2 \\ 1 & 4 \end{vmatrix} - 2 \begin{vmatrix} 3 & 2 \\ 1 & 4 \end{vmatrix} + 7 \begin{vmatrix} 3 & -5 \\ 1 & 1 \end{vmatrix}$$

$$= [1 \times (-22)] - [2 \times 10] + [7 \times 8]$$
$$= -22 - 20 + 56$$
$$= 14$$

You are now in a position to find the inverse of a 3×3 matrix. So consider, once again, the matrix

$$\begin{pmatrix} 1 & 2 & 4 \\ -1 & 3 & 0 \\ 0 & 1 & 5 \end{pmatrix}$$

The first thing to do is to take each element of the matrix in turn and replace it by its **minor**. If you take an element and cross out the row and the column in which it lies, the determinant of what is left is called the minor of that element.

To find the minor of the '3' you cross out its row and column:

$$\begin{pmatrix} 1 & 2 & 4 \\ -1 & 3 & 0 \\ 0 & 1 & 5 \end{pmatrix}$$

The minor of 3 is $\begin{vmatrix} 1 & 4 \\ 0 & 5 \end{vmatrix} = 5.$

The minor of the '0' in the bottom left-hand corner is

$$\begin{vmatrix} 2 & 4 \\ 3 & 0 \end{vmatrix} = -12$$

The minor of the '−1' is

$$\begin{vmatrix} 2 & 4 \\ 1 & 5 \end{vmatrix} = 6$$

so the matrix of minors is

$$\begin{pmatrix} 15 & -5 & -1 \\ 6 & 5 & 1 \\ -12 & 4 & 5 \end{pmatrix}$$

You must now change some of the signs according to the **alternating law of signs**:

$$\begin{pmatrix} + & - & + \\ - & + & - \\ + & - & + \end{pmatrix}$$

So, wherever there is a minus sign in the above, you must change the sign of the corresponding element in the matrix of minors. Thus you get

$$\begin{pmatrix} 15 & 5 & -1 \\ -6 & 5 & -1 \\ -12 & -4 & 5 \end{pmatrix}$$

This is called the **matrix of cofactors**.

Now transpose the matrix of cofactors:

$$\begin{pmatrix} 15 & -6 & -12 \\ 5 & 5 & -4 \\ -1 & -1 & 5 \end{pmatrix}$$

Finally, as with 2×2 matrices, you divide by the determinant. You will remember from page 68 that the determinant is 21. So the inverse of

$$\begin{pmatrix} 1 & 2 & 4 \\ -1 & 3 & 0 \\ 0 & 1 & 5 \end{pmatrix} \quad \text{is} \quad \tfrac{1}{21}\begin{pmatrix} 15 & -6 & -12 \\ 5 & 5 & -4 \\ -1 & -1 & 5 \end{pmatrix}$$

To check this, multiply the two together:

$$\tfrac{1}{21}\begin{pmatrix} 15 & -6 & -12 \\ 5 & 5 & -4 \\ -1 & -1 & 5 \end{pmatrix}\begin{pmatrix} 1 & 2 & 4 \\ -1 & 3 & 0 \\ 0 & 1 & 5 \end{pmatrix} = \tfrac{1}{21}\begin{pmatrix} 21 & 0 & 0 \\ 0 & 21 & 0 \\ 0 & 0 & 21 \end{pmatrix}$$

$$= \begin{pmatrix} 1 & 0 & 0 \\ 0 & 1 & 0 \\ 0 & 0 & 1 \end{pmatrix} \text{ as required}$$

Example 8

Find the inverse of

$$\begin{pmatrix} 1 & 2 & 7 \\ 3 & -5 & 2 \\ 1 & 1 & 4 \end{pmatrix}$$

The first thing to do is to calculate the determinant. From example 7 on page 69 you know that this is 14.

The matrix of minors is

$$\begin{pmatrix} -22 & 10 & 8 \\ 1 & -3 & -1 \\ 39 & -19 & -11 \end{pmatrix}$$

The matrix of cofactors is

$$\begin{pmatrix} -22 & -10 & 8 \\ -1 & -3 & 1 \\ 39 & 19 & -11 \end{pmatrix}$$

The transpose of this is

$$\begin{pmatrix} -22 & -1 & 39 \\ -10 & -3 & 19 \\ 8 & 1 & -11 \end{pmatrix}$$

So the inverse is

$$\frac{1}{14}\begin{pmatrix} -22 & -1 & 39 \\ -10 & -3 & 19 \\ 8 & 1 & -11 \end{pmatrix}$$

Check this by calculating the product:

$$\frac{1}{14}\begin{pmatrix} -22 & -1 & 39 \\ -10 & -3 & 19 \\ 8 & 1 & -11 \end{pmatrix}\begin{pmatrix} 1 & 2 & 7 \\ 3 & -5 & 2 \\ 1 & 1 & 4 \end{pmatrix} = \frac{1}{14}\begin{pmatrix} 14 & 0 & 0 \\ 0 & 14 & 0 \\ 0 & 0 & 14 \end{pmatrix}$$

$$= \begin{pmatrix} 1 & 0 & 0 \\ 0 & 1 & 0 \\ 0 & 0 & 1 \end{pmatrix} \text{ as required}$$

Because the last step in finding an inverse matrix is to divide by the determinant, the matrix will have no inverse if the determinant is zero.

As you saw on page 65, a matrix whose determinant is zero is called a *singular* matrix and it has no inverse. The only matrices which have inverses are non-singular square matrices.

Exercise 3B

1 Evaluate:

(a) $\begin{pmatrix} 1 & -2 \\ 7 & 5 \end{pmatrix} + \begin{pmatrix} -8 & 9 \\ 3 & -2 \end{pmatrix}$

(b) $\begin{pmatrix} -7 & 6 \\ -5 & 4 \end{pmatrix} + \begin{pmatrix} -9 & 2 \\ 1 & -3 \end{pmatrix}$

(c) $\begin{pmatrix} 10 & -7 \\ -5 & 4 \end{pmatrix} - \begin{pmatrix} -9 & 3 \\ 4 & -9 \end{pmatrix}$

(d) $\begin{pmatrix} -8 & 3 \\ 5 & -11 \end{pmatrix} - \begin{pmatrix} 4 & -9 \\ 8 & -20 \end{pmatrix}$

(e) $\begin{pmatrix} 5 & -2 & -10 \\ -15 & 2 & 27 \\ 3 & -5 & 9 \end{pmatrix} + \begin{pmatrix} 27 & 6 & -13 \\ 4 & -8 & 21 \\ 3 & -15 & 36 \end{pmatrix}$

(f) $\begin{pmatrix} 52 & -7 & -39 \\ 4 & 8 & 19 \\ 21 & -3 & -10 \end{pmatrix} - \begin{pmatrix} -15 & 4 & -13 \\ 7 & 18 & 24 \\ -10 & -15 & 21 \end{pmatrix}$

2 Evaluate:

(a) $\begin{pmatrix} 2 & 1 \\ -3 & 2 \end{pmatrix} \begin{pmatrix} -1 & 4 \\ 0 & 2 \end{pmatrix}$

(b) $\begin{pmatrix} -1 & 4 \\ 0 & 2 \end{pmatrix} \begin{pmatrix} 2 & 1 \\ -3 & 2 \end{pmatrix}$

(c) $\begin{pmatrix} 1 & -3 & 2 \\ -1 & 0 & 4 \\ 0 & 0 & -2 \end{pmatrix} \begin{pmatrix} 5 \\ 1 \\ 3 \end{pmatrix}$

(d) $\begin{pmatrix} 3 & -5 & 4 \\ 0 & -6 & 3 \end{pmatrix} \begin{pmatrix} 1 & -3 & 2 \\ -1 & 0 & 4 \\ 0 & 0 & -2 \end{pmatrix}$

(e) $\begin{pmatrix} 1 & 5 & -4 \\ 4 & -1 & 3 \\ 0 & 0 & -2 \end{pmatrix} \begin{pmatrix} 6 & 1 & 5 \\ -2 & 0 & 3 \\ 1 & 1 & -2 \end{pmatrix}$

(f) $\begin{pmatrix} 2 & -1 & 6 \\ 3 & 5 & -2 \\ -4 & 6 & -2 \end{pmatrix} \begin{pmatrix} 8 & 1 & 7 \\ 2 & -3 & 1 \\ 1 & 1 & -2 \end{pmatrix}$

(g) $\begin{pmatrix} 2 & -3 & 1 \\ 1 & -2 & 4 \\ 5 & 4 & -3 \end{pmatrix} \begin{pmatrix} 2 & 3 & -7 \\ 6 & 3 & -1 \\ -1 & 2 & 3 \end{pmatrix}$

(h) $\begin{pmatrix} 8 & 1 & 7 \\ 2 & -3 & 1 \\ 1 & 1 & -2 \end{pmatrix} \begin{pmatrix} 2 & -1 & 6 \\ 3 & 5 & -2 \\ -4 & 6 & -2 \end{pmatrix}$

(i) $\begin{pmatrix} 6 & 1 & 5 \\ -2 & 0 & 3 \\ 1 & 1 & -2 \end{pmatrix} \begin{pmatrix} 1 & 5 & -4 \\ 4 & -1 & 3 \\ 0 & 0 & -2 \end{pmatrix}$

3 Evaluate:

(a) $\begin{vmatrix} 1 & 2 \\ 1 & 3 \end{vmatrix}$

(b) $\begin{vmatrix} 1 & -1 \\ 2 & -4 \end{vmatrix}$

(c) $\begin{vmatrix} 2 & 1 \\ 3 & -1 \end{vmatrix}$

(d) $\begin{vmatrix} 0 & -2 \\ 1 & -4 \end{vmatrix}$

(e) $\begin{vmatrix} 1 & 2 & -3 \\ 1 & 1 & 0 \\ -1 & 4 & -6 \end{vmatrix}$

(f) $\begin{vmatrix} -2 & 7 & 3 \\ 1 & 2 & 4 \\ -1 & 2 & 0 \end{vmatrix}$

(g) $\begin{vmatrix} 2 & -3 & 6 \\ -2 & 4 & 5 \\ -1 & 0 & -5 \end{vmatrix}$

(h) $\begin{vmatrix} 1 & 2 & -3 \\ 2 & 2 & -4 \\ -4 & 2 & 1 \end{vmatrix}$

(i) $\begin{vmatrix} -3 & 1 & 2 \\ 0 & 1 & -3 \\ 3 & 3 & -2 \end{vmatrix}$
(j) $\begin{vmatrix} 3 & 4 & 1 \\ 4 & 1 & 3 \\ 2 & 5 & 2 \end{vmatrix}$

(k) $\begin{vmatrix} 1 & -1 & 3 \\ 2 & -2 & 4 \\ 3 & -3 & 5 \end{vmatrix}$
(l) $\begin{vmatrix} 0 & -1 & 2 \\ 1 & 0 & 7 \\ -2 & -7 & 0 \end{vmatrix}$

(m) $\begin{vmatrix} 0 & -1 & 2 \\ 1 & 0 & -2 \\ -2 & 2 & 0 \end{vmatrix}$
(n) $\begin{vmatrix} 3 & 0 & 2 \\ 0 & -1 & 4 \\ 1 & 1 & -2 \end{vmatrix}$

(o) $\begin{vmatrix} 2 & 0 & 1 \\ -1 & 2 & 3 \\ 1 & 0 & 2 \end{vmatrix}$
(p) $\begin{vmatrix} 5 & 1 & 3 \\ -2 & 0 & 1 \\ 1 & 1 & -2 \end{vmatrix}$

4 Find the inverse of:

(a) $\begin{pmatrix} 2 & 5 \\ -1 & 4 \end{pmatrix}$
(b) $\begin{pmatrix} -3 & 2 \\ -1 & 7 \end{pmatrix}$

(c) $\begin{pmatrix} 2 & -3 \\ 1 & -4 \end{pmatrix}$
(d) $\begin{pmatrix} 0 & 1 \\ -3 & 2 \end{pmatrix}$

(e) $\begin{pmatrix} -3 & 7 \\ 9 & 22 \end{pmatrix}$
(f) $\begin{pmatrix} 2 & 1 & -5 \\ 1 & 0 & -2 \\ 0 & 0 & 3 \end{pmatrix}$

(g) $\begin{pmatrix} -1 & 2 & 1 \\ 0 & 0 & -2 \\ 1 & -5 & 4 \end{pmatrix}$
(h) $\begin{pmatrix} -1 & 2 & 3 \\ 1 & 1 & 2 \\ 5 & -1 & 4 \end{pmatrix}$

(i) $\begin{pmatrix} -2 & 3 & -4 \\ 1 & 2 & -3 \\ -3 & 0 & -2 \end{pmatrix}$
(j) $\begin{pmatrix} 2 & -1 & 2 \\ 1 & -1 & 1 \\ 2 & 1 & -3 \end{pmatrix}$

(k) $\begin{pmatrix} 3 & 2 & -6 \\ 1 & 1 & -2 \\ 2 & 2 & -1 \end{pmatrix}$
(l) $\begin{pmatrix} 4 & -5 & 2 \\ 0 & 1 & -7 \\ 1 & 1 & -2 \end{pmatrix}$

(m) $\begin{pmatrix} 3 & 2 & -3 \\ 1 & 1 & -4 \\ 2 & 2 & -6 \end{pmatrix}$
(n) $\begin{pmatrix} -2 & 1 & -2 \\ 4 & 3 & 1 \\ 0 & 1 & -6 \end{pmatrix}$

(o) $\begin{pmatrix} 2 & 1 & -3 \\ 1 & -1 & 2 \\ 3 & 2 & 1 \end{pmatrix}$
(p) $\begin{pmatrix} 2 & 2 & 1 \\ 4 & 1 & 5 \\ -1 & 1 & 7 \end{pmatrix}$

3.6 Using a matrix to represent a linear transformation

You should remember from the work you have done on vectors that in two dimensions

$$\mathbf{i} = \begin{pmatrix} 1 \\ 0 \end{pmatrix} \quad \text{and} \quad \mathbf{j} = \begin{pmatrix} 0 \\ 1 \end{pmatrix}$$

and that in three dimensions

$$\mathbf{i} = \begin{pmatrix} 1 \\ 0 \\ 0 \end{pmatrix}, \mathbf{j} = \begin{pmatrix} 0 \\ 1 \\ 0 \end{pmatrix} \quad \text{and} \quad \mathbf{k} = \begin{pmatrix} 0 \\ 0 \\ 1 \end{pmatrix}$$

Now if $T : \begin{pmatrix} x \\ y \\ z \end{pmatrix} \mapsto \begin{pmatrix} x+y \\ y+z \\ x+z \end{pmatrix}$ then:

$$T\begin{pmatrix} 1 \\ 0 \\ 0 \end{pmatrix} = \begin{pmatrix} 1+0 \\ 0+0 \\ 1+0 \end{pmatrix} = \begin{pmatrix} 1 \\ 0 \\ 1 \end{pmatrix}$$

$$T\begin{pmatrix} 0 \\ 1 \\ 0 \end{pmatrix} = \begin{pmatrix} 0+1 \\ 1+0 \\ 0+0 \end{pmatrix} = \begin{pmatrix} 1 \\ 1 \\ 0 \end{pmatrix}$$

$$T\begin{pmatrix} 0 \\ 0 \\ 1 \end{pmatrix} = \begin{pmatrix} 0+0 \\ 0+1 \\ 0+1 \end{pmatrix} = \begin{pmatrix} 0 \\ 1 \\ 1 \end{pmatrix}$$

If you write the images of $\mathbf{i}$, $\mathbf{j}$ and $\mathbf{k}$ as the columns of a 3×3 matrix, you get

$$\begin{pmatrix} 1 & 1 & 0 \\ 0 & 1 & 1 \\ 1 & 0 & 1 \end{pmatrix}$$

and

$$\begin{pmatrix} 1 & 1 & 0 \\ 0 & 1 & 1 \\ 1 & 0 & 1 \end{pmatrix} \begin{pmatrix} x \\ y \\ z \end{pmatrix} = \begin{pmatrix} x+y \\ y+z \\ x+z \end{pmatrix}$$

That is, the matrix $\begin{pmatrix} 1 & 1 & 0 \\ 0 & 1 & 1 \\ 1 & 0 & 1 \end{pmatrix}$ maps $\begin{pmatrix} x \\ y \\ z \end{pmatrix}$ to $\begin{pmatrix} x+y \\ y+z \\ x+z \end{pmatrix}$.

But
$$T\begin{pmatrix} x \\ y \\ z \end{pmatrix} = \begin{pmatrix} x+y \\ y+z \\ x+z \end{pmatrix}$$

So the matrix has exactly the same effect on $\begin{pmatrix} x \\ y \\ z \end{pmatrix}$ as does the

transformation T. So you can say that the matrix $\begin{pmatrix} 1 & 1 & 0 \\ 0 & 1 & 1 \\ 1 & 0 & 1 \end{pmatrix}$

represents T.

■ **To obtain a matrix which represents a linear transformation T, find the images of i and j (in two dimensions) or of i, j and k (in three dimensions) and write these as the columns of a square 2×2 or 3×3 matrix.**

Example 9
Find the matrix which represents the linear transformation

$$T: \begin{pmatrix} x \\ y \end{pmatrix} \mapsto \begin{pmatrix} x \\ -y \end{pmatrix}$$

$$T\begin{pmatrix} 1 \\ 0 \end{pmatrix} = \begin{pmatrix} 1 \\ 0 \end{pmatrix}$$

$$T\begin{pmatrix} 0 \\ 1 \end{pmatrix} = \begin{pmatrix} 0 \\ -1 \end{pmatrix}$$

So the matrix that represents T is $\begin{pmatrix} 1 & 0 \\ 0 & -1 \end{pmatrix}$.

Example 10
Find the matrix which represents the linear transformation

$$T: \begin{pmatrix} x \\ y \\ z \end{pmatrix} \mapsto 6\begin{pmatrix} x \\ x-y \\ x+z \end{pmatrix}$$

$$T\begin{pmatrix} 1 \\ 0 \\ 0 \end{pmatrix} = \begin{pmatrix} 6 \\ 6 \\ 6 \end{pmatrix}, \quad T\begin{pmatrix} 0 \\ 1 \\ 0 \end{pmatrix} = \begin{pmatrix} 0 \\ -6 \\ 0 \end{pmatrix} \quad \text{and} \quad T\begin{pmatrix} 0 \\ 0 \\ 1 \end{pmatrix} = \begin{pmatrix} 0 \\ 0 \\ 6 \end{pmatrix}$$

So the matrix is $\begin{pmatrix} 6 & 0 & 0 \\ 6 & -6 & 0 \\ 6 & 0 & 6 \end{pmatrix}$.

3.7 Using a matrix to represent a combination of linear transformations

Suppose you have two linear transformations:

$$T: \begin{pmatrix} x \\ y \\ z \end{pmatrix} \mapsto \begin{pmatrix} x \\ x+y \\ x+z \end{pmatrix}$$

and

$$V: \begin{pmatrix} x \\ y \\ z \end{pmatrix} \mapsto \begin{pmatrix} x+y \\ y \\ y+z \end{pmatrix}$$

Then

$$TV \begin{pmatrix} x \\ y \\ z \end{pmatrix} = T \begin{pmatrix} x+y \\ y \\ y+z \end{pmatrix}$$

$$= \begin{pmatrix} x+y \\ x+2y \\ x+2y+z \end{pmatrix}$$

Now

$$T \begin{pmatrix} 1 \\ 0 \\ 0 \end{pmatrix} = \begin{pmatrix} 1 \\ 1 \\ 1 \end{pmatrix}, \quad T \begin{pmatrix} 0 \\ 1 \\ 0 \end{pmatrix} = \begin{pmatrix} 0 \\ 1 \\ 0 \end{pmatrix} \quad \text{and} \quad T \begin{pmatrix} 0 \\ 0 \\ 1 \end{pmatrix} = \begin{pmatrix} 0 \\ 0 \\ 1 \end{pmatrix}$$

So $\begin{pmatrix} 1 & 0 & 0 \\ 1 & 1 & 0 \\ 1 & 0 & 1 \end{pmatrix}$ represents T.

Also:

$$V \begin{pmatrix} 1 \\ 0 \\ 0 \end{pmatrix} = \begin{pmatrix} 1 \\ 0 \\ 0 \end{pmatrix}, \quad V \begin{pmatrix} 0 \\ 1 \\ 0 \end{pmatrix} = \begin{pmatrix} 1 \\ 1 \\ 1 \end{pmatrix} \quad \text{and} \quad V \begin{pmatrix} 0 \\ 0 \\ 1 \end{pmatrix} = \begin{pmatrix} 0 \\ 0 \\ 1 \end{pmatrix}$$

So $\begin{pmatrix} 1 & 1 & 0 \\ 0 & 1 & 0 \\ 0 & 1 & 1 \end{pmatrix}$ represents V.

Again

$$TV\begin{pmatrix}1\\0\\0\end{pmatrix}=\begin{pmatrix}1\\1\\1\end{pmatrix}, \quad TV\begin{pmatrix}0\\1\\0\end{pmatrix}=\begin{pmatrix}1\\2\\2\end{pmatrix} \quad \text{and} \quad TV\begin{pmatrix}0\\0\\1\end{pmatrix}=\begin{pmatrix}0\\0\\1\end{pmatrix}$$

So $\begin{pmatrix}1&1&0\\1&2&0\\1&2&1\end{pmatrix}$ represents TV.

Now $\begin{pmatrix}1&0&0\\1&1&0\\1&0&1\end{pmatrix}\begin{pmatrix}1&1&0\\0&1&0\\0&1&1\end{pmatrix}=\begin{pmatrix}1&1&0\\1&2&0\\1&2&1\end{pmatrix}$

That is:

(matrix representing T) × (matrix representing V) = matrix representing TV

You must remember, however, that if **T** is the matrix representing the linear transformation T and **V** is the matrix representing the linear transformation V then the matrix **TV** represents the combined transformation TV, where you do V first and then T. The matrix **VT** represents the linear transformation that results from doing T first then V, and it is, in general, as you know, *not* the same as **TV**.

3.8 Using a matrix to represent an inverse transformation

In Example 3 (page 58) you saw that if

$$T:\begin{pmatrix}x\\y\\z\end{pmatrix}\mapsto\begin{pmatrix}x\\x+y\\z\end{pmatrix}$$

then:

$$T^{-1}:\begin{pmatrix}x\\y\\z\end{pmatrix}\mapsto\begin{pmatrix}x\\y-x\\z\end{pmatrix}$$

Now

$$T\begin{pmatrix}1\\0\\0\end{pmatrix}=\begin{pmatrix}1\\1\\0\end{pmatrix}, \quad T\begin{pmatrix}0\\1\\0\end{pmatrix}=\begin{pmatrix}0\\1\\0\end{pmatrix} \quad \text{and} \quad T\begin{pmatrix}0\\0\\1\end{pmatrix}=\begin{pmatrix}0\\0\\1\end{pmatrix}$$

So $\begin{pmatrix}1&0&0\\1&1&0\\0&0&1\end{pmatrix}$ represents T.

Also:

$$T^{-1}\begin{pmatrix} 1 \\ 0 \\ 0 \end{pmatrix} = \begin{pmatrix} 1 \\ -1 \\ 0 \end{pmatrix}, \quad T^{-1}\begin{pmatrix} 0 \\ 1 \\ 0 \end{pmatrix} = \begin{pmatrix} 0 \\ 1 \\ 0 \end{pmatrix} \quad \text{and} \quad T^{-1}\begin{pmatrix} 0 \\ 0 \\ 1 \end{pmatrix} = \begin{pmatrix} 0 \\ 0 \\ 1 \end{pmatrix}$$

So $\begin{pmatrix} 1 & 0 & 0 \\ -1 & 1 & 0 \\ 0 & 0 & 1 \end{pmatrix}$ represents T^{-1}.

But $\begin{pmatrix} 1 & 0 & 0 \\ 1 & 1 & 0 \\ 0 & 0 & 1 \end{pmatrix} \begin{pmatrix} 1 & 0 & 0 \\ -1 & 1 & 0 \\ 0 & 0 & 1 \end{pmatrix} = \begin{pmatrix} 1 & 0 & 0 \\ 0 & 1 & 0 \\ 0 & 0 & 1 \end{pmatrix} = \mathbf{I}$

That is:

$\begin{pmatrix} 1 & 0 & 0 \\ -1 & 1 & 0 \\ 0 & 0 & 1 \end{pmatrix}$ is the inverse of $\begin{pmatrix} 1 & 0 & 0 \\ 1 & 1 & 0 \\ 0 & 0 & 1 \end{pmatrix}$

So if the matrix $\mathbf{T}$ represents the linear transformation T, then the matrix $\mathbf{T}^{-1}$ represents the linear transformation T^{-1}, if it exists.

3.9 Using a matrix to represent the inverse of a combination of transformations

If
$$T: \begin{pmatrix} x \\ y \\ z \end{pmatrix} \mapsto \begin{pmatrix} x \\ x+y \\ x+z \end{pmatrix}$$

and
$$V: \begin{pmatrix} x \\ y \\ z \end{pmatrix} \mapsto \begin{pmatrix} x+y \\ y \\ y+z \end{pmatrix}$$

then you saw in section 3.7 that

$$TV: \begin{pmatrix} x \\ y \\ z \end{pmatrix} \mapsto \begin{pmatrix} x+y \\ x+2y \\ x+2y+z \end{pmatrix}$$

If
$$\begin{pmatrix} p_1 \\ p_2 \\ p_3 \end{pmatrix} = \begin{pmatrix} x+y \\ x+2y \\ x+2y+z \end{pmatrix}$$

then
$$z = p_3 - p_2$$
$$y = p_2 - p_1$$

and
$$x = p_1 - y = p_1 - p_2 + p_1 = 2p_1 - p_2$$

So
$$(TV)^{-1} : \begin{pmatrix} x \\ y \\ z \end{pmatrix} \mapsto \begin{pmatrix} 2x - y \\ y - x \\ z - y \end{pmatrix}$$

Now:

$$(TV)^{-1} \begin{pmatrix} 1 \\ 0 \\ 0 \end{pmatrix} = \begin{pmatrix} 2 \\ -1 \\ 0 \end{pmatrix}, \quad (TV)^{-1} \begin{pmatrix} 0 \\ 1 \\ 0 \end{pmatrix} = \begin{pmatrix} -1 \\ 1 \\ -1 \end{pmatrix} \quad \text{and} \quad (TV)^{-1} \begin{pmatrix} 0 \\ 0 \\ 1 \end{pmatrix} = \begin{pmatrix} 0 \\ 0 \\ 1 \end{pmatrix}$$

Hence $(\mathbf{TV})^{-1} = \begin{pmatrix} 2 & -1 & 0 \\ -1 & 1 & 0 \\ 0 & -1 & 1 \end{pmatrix}$ represents $(TV)^{-1}$.

Now $\mathbf{T} = \begin{pmatrix} 1 & 0 & 0 \\ 1 & 1 & 0 \\ 1 & 0 & 1 \end{pmatrix}$ represents T and you can easily find

$$\mathbf{T}^{-1} = \begin{pmatrix} 1 & 0 & 0 \\ -1 & 1 & 0 \\ -1 & 0 & 1 \end{pmatrix}$$

Also $\mathbf{V} = \begin{pmatrix} 1 & 1 & 0 \\ 0 & 1 & 0 \\ 0 & 1 & 1 \end{pmatrix}$ represents V and you can easily find

$$\mathbf{V}^{-1} = \begin{pmatrix} 1 & -1 & 0 \\ 0 & 1 & 0 \\ 0 & -1 & 1 \end{pmatrix}$$

Now :
$$\mathbf{V}^{-1}\mathbf{T}^{-1} = \begin{pmatrix} 1 & -1 & 0 \\ 0 & 1 & 0 \\ 0 & -1 & 1 \end{pmatrix} \begin{pmatrix} 1 & 0 & 0 \\ -1 & 1 & 0 \\ -1 & 0 & 1 \end{pmatrix}$$

$$= \begin{pmatrix} 2 & -1 & 0 \\ -1 & 1 & 0 \\ 0 & -1 & 1 \end{pmatrix} = (\mathbf{TV})^{-1}$$

That is, you can find the matrix representing the inverse of the combined transformation TV by multiplying $\mathbf{V}^{-1}\mathbf{T}^{-1}$, the matrix representing V^{-1} by the matrix representing T^{-1}, *in that order*. This result is true for non-singular square matrices in general. That is:

$$(\mathbf{AB})^{-1} = \mathbf{B}^{-1}\mathbf{A}^{-1}$$

A similar result holds for the transposes of matrices.

If $\mathbf{A} = \begin{pmatrix} 1 & 2 & 1 \\ 0 & -1 & 2 \\ 1 & 1 & 3 \end{pmatrix}$ and $\mathbf{B} = \begin{pmatrix} 2 & 1 & 3 \\ -1 & 4 & 1 \\ -1 & 0 & -1 \end{pmatrix}$

then $\quad \mathbf{AB} = \begin{pmatrix} 1 & 2 & 1 \\ 0 & -1 & 2 \\ 1 & 1 & 3 \end{pmatrix} \begin{pmatrix} 2 & 1 & 3 \\ -1 & 4 & 1 \\ -1 & 0 & -1 \end{pmatrix}$

$\qquad = \begin{pmatrix} -1 & 9 & 4 \\ -1 & -4 & -3 \\ -2 & 5 & 1 \end{pmatrix}$

So $\qquad (\mathbf{AB})^{\mathrm{T}} = \begin{pmatrix} -1 & -1 & -2 \\ 9 & -4 & 5 \\ 4 & -3 & 1 \end{pmatrix}$

Now $\qquad \mathbf{A}^{\mathrm{T}} = \begin{pmatrix} 1 & 0 & 1 \\ 2 & -1 & 1 \\ 1 & 2 & 3 \end{pmatrix}$

and $\qquad \mathbf{B}^{\mathrm{T}} = \begin{pmatrix} 2 & -1 & -1 \\ 1 & 4 & 0 \\ 3 & 1 & -1 \end{pmatrix}$

so $\mathbf{B}^{\mathrm{T}}\mathbf{A}^{\mathrm{T}} = \begin{pmatrix} 2 & -1 & -1 \\ 1 & 4 & 0 \\ 3 & 1 & -1 \end{pmatrix} \begin{pmatrix} 1 & 0 & 1 \\ 2 & -1 & 1 \\ 1 & 2 & 3 \end{pmatrix} = \begin{pmatrix} -1 & -1 & -2 \\ 9 & -4 & 5 \\ 4 & -3 & 1 \end{pmatrix}$

$\qquad = (\mathbf{AB})^{\mathrm{T}}$

That is
- $(\mathbf{AB})^{\mathrm{T}} = \mathbf{B}^{\mathrm{T}}\mathbf{A}^{\mathrm{T}}$ **and this result is generally true for matrices which commute.**

Example 11

Given that

$$\mathbf{A} = \begin{pmatrix} 1 & 0 & 2 \\ -1 & 2 & 1 \\ 1 & 1 & 2 \end{pmatrix} \quad \text{and} \quad \mathbf{B} = \begin{pmatrix} 0 & -1 & 2 \\ 1 & 1 & 2 \\ -1 & 3 & 1 \end{pmatrix}$$

(a) find $\mathbf{A}^{-1}$ and $\mathbf{B}^{-1}$
(b) find $\mathbf{A}^{\mathrm{T}}$ and $\mathbf{B}^{\mathrm{T}}$.
Hence find
(c) $(\mathbf{AB})^{-1}$ and $(\mathbf{BA})^{-1}$
(d) $(\mathbf{AB})^{\mathrm{T}}$ and $(\mathbf{BA})^{\mathrm{T}}$.

(a)
$$\det \mathbf{A} = \begin{vmatrix} 1 & 0 & 2 \\ -1 & 2 & 1 \\ 1 & 1 & 2 \end{vmatrix}$$

$$= 1 \begin{vmatrix} 2 & 1 \\ 1 & 2 \end{vmatrix} + 2 \begin{vmatrix} -1 & 2 \\ 1 & 1 \end{vmatrix}$$

$$= 3 + 2(-3) = -3$$

For **A** the matrix of minors is

$$\begin{pmatrix} 3 & -3 & -3 \\ -2 & 0 & 1 \\ -4 & 3 & 2 \end{pmatrix}$$

The matrix of cofactors is

$$\begin{pmatrix} 3 & 3 & -3 \\ 2 & 0 & -1 \\ -4 & -3 & 2 \end{pmatrix}$$

The transpose is

$$\begin{pmatrix} 3 & 2 & -4 \\ 3 & 0 & -3 \\ -3 & -1 & 2 \end{pmatrix}$$

So $\mathbf{A}^{-1} = -\frac{1}{3} \begin{pmatrix} 3 & 2 & -4 \\ 3 & 0 & -3 \\ -3 & -1 & 2 \end{pmatrix}$

$$\det \mathbf{B} = \begin{vmatrix} 0 & -1 & 2 \\ 1 & 1 & 2 \\ -1 & 3 & 1 \end{vmatrix}$$

$$= 1 \begin{vmatrix} 1 & 2 \\ -1 & 1 \end{vmatrix} + 2 \begin{vmatrix} 1 & 1 \\ -1 & 3 \end{vmatrix}$$

$$= 3 + 2(4) = 11$$

For **B** the matrix of minors is

$$\begin{pmatrix} -5 & 3 & 4 \\ -7 & 2 & -1 \\ -4 & -2 & 1 \end{pmatrix}$$

The matrix of cofactors is

$$\begin{pmatrix} -5 & -3 & 4 \\ 7 & 2 & 1 \\ -4 & 2 & 1 \end{pmatrix}$$

The transpose is

$$\begin{pmatrix} -5 & 7 & -4 \\ -3 & 2 & 2 \\ 4 & 1 & 1 \end{pmatrix}$$

So $\mathbf{B}^{-1} = \frac{1}{11}\begin{pmatrix} -5 & 7 & -4 \\ -3 & 2 & 2 \\ 4 & 1 & 1 \end{pmatrix}$

(b) $\mathbf{A}^{\mathrm{T}} = \begin{pmatrix} 1 & -1 & 1 \\ 0 & 2 & 1 \\ 2 & 1 & 2 \end{pmatrix}$

$\mathbf{B}^{\mathrm{T}} = \begin{pmatrix} 0 & 1 & -1 \\ -1 & 1 & 3 \\ 2 & 2 & 1 \end{pmatrix}$

(c) $(\mathbf{AB})^{-1} = \mathbf{B}^{-1}\,\mathbf{A}^{-1}$

$$= \frac{1}{11}\begin{pmatrix} -5 & 7 & -4 \\ -3 & 2 & 2 \\ 4 & 1 & 1 \end{pmatrix} \times -\frac{1}{3}\begin{pmatrix} 3 & 2 & -4 \\ 3 & 0 & -3 \\ -3 & -1 & 2 \end{pmatrix}$$

$$= -\frac{1}{33}\begin{pmatrix} 18 & -6 & -9 \\ -9 & -8 & 10 \\ 12 & 7 & -17 \end{pmatrix}$$

$(\mathbf{BA})^{-1} = \mathbf{A}^{-1}\,\mathbf{B}^{-1}$

$$= -\frac{1}{3}\begin{pmatrix} 3 & 2 & -4 \\ 3 & 0 & -3 \\ -3 & -1 & 2 \end{pmatrix} \times \frac{1}{11}\begin{pmatrix} -5 & 7 & -4 \\ -3 & 2 & 2 \\ 4 & 1 & 1 \end{pmatrix}$$

$$= -\frac{1}{33}\begin{pmatrix} -37 & 21 & -12 \\ -27 & 18 & -15 \\ 26 & -21 & 12 \end{pmatrix}$$

(d) $(\mathbf{AB})^{\mathrm{T}} = \mathbf{B}^{\mathrm{T}}\,\mathbf{A}^{\mathrm{T}}$

$$= \begin{pmatrix} 0 & 1 & -1 \\ -1 & 1 & 3 \\ 2 & 2 & 1 \end{pmatrix}\begin{pmatrix} 1 & -1 & 1 \\ 0 & 2 & 1 \\ 2 & 1 & 2 \end{pmatrix}$$

$$= \begin{pmatrix} -2 & 1 & -1 \\ 5 & 6 & 6 \\ 4 & 3 & 6 \end{pmatrix}$$

$(\mathbf{BA})^{\mathrm{T}} = \mathbf{A}^{\mathrm{T}} \mathbf{B}^{\mathrm{T}}$

$$= \begin{pmatrix} 1 & -1 & 1 \\ 0 & 2 & 1 \\ 2 & 1 & 2 \end{pmatrix} \begin{pmatrix} 0 & 1 & -1 \\ -1 & 1 & 3 \\ 2 & 2 & 1 \end{pmatrix}$$

$$= \begin{pmatrix} 3 & 2 & -3 \\ 0 & 4 & 7 \\ 3 & 7 & 3 \end{pmatrix}$$

Exercise 3C

1 Given that

$$T : \begin{pmatrix} x \\ y \end{pmatrix} \mapsto \begin{pmatrix} x + y \\ x - y \end{pmatrix}$$

$$U : \begin{pmatrix} x \\ y \end{pmatrix} \mapsto \begin{pmatrix} x \\ 2x + y \end{pmatrix}$$

find the matrix that represents (a) T (b) U (c) TU (d) UT.

2 Given that

$$T : \begin{pmatrix} x \\ y \end{pmatrix} \mapsto \begin{pmatrix} x + y \\ x - y \end{pmatrix}$$

$$U : \begin{pmatrix} x \\ y \end{pmatrix} \mapsto \begin{pmatrix} x \\ 2x + y \end{pmatrix}$$

find the matrix that represents (a) T^{-1} (b) U^{-1} (c) $(TU)^{-1}$.

3 Given that

$$T : \begin{pmatrix} x \\ y \end{pmatrix} \mapsto \begin{pmatrix} y \\ x \end{pmatrix}$$

$$U : \begin{pmatrix} x \\ y \end{pmatrix} \mapsto \begin{pmatrix} 2x \\ y \end{pmatrix}$$

find the matrix that represents (a) T (b) U (c) TU (d) UT.

4 Given that

$$T : \begin{pmatrix} x \\ y \end{pmatrix} \mapsto \begin{pmatrix} y \\ x \end{pmatrix}$$

$$U : \begin{pmatrix} x \\ y \end{pmatrix} \mapsto \begin{pmatrix} 2x \\ y \end{pmatrix}$$

find the matrix that represents (a) T^{-1} (b) U^{-1} (c) $(UT)^{-1}$.

5 Given that

$$T: \begin{pmatrix} x \\ y \\ z \end{pmatrix} \mapsto \begin{pmatrix} 2x + z \\ y \\ -y + z \end{pmatrix}$$

$$U: \begin{pmatrix} x \\ y \\ z \end{pmatrix} \mapsto \begin{pmatrix} -x + 2y - 3z \\ 2x - y + 4z \\ 3x + 4y + z \end{pmatrix}$$

find the matrix that represents (a) T (b) U (c) TU (d) UT.

6 Given that

$$T: \begin{pmatrix} x \\ y \\ z \end{pmatrix} \mapsto \begin{pmatrix} 2x + z \\ y \\ -y + z \end{pmatrix}$$

$$U: \begin{pmatrix} x \\ y \\ z \end{pmatrix} \mapsto \begin{pmatrix} -x + 2y - 3z \\ 2x - y + 4z \\ 3x + 4y + z \end{pmatrix}$$

find the matrix that represents
(a) T^{-1} (b) U^{-1} (c) $(TU)^{-1}$ (d) $(UT)^{-1}$.

7 Given that

$$T: \begin{pmatrix} x \\ y \\ z \end{pmatrix} \mapsto \begin{pmatrix} x - y + 3z \\ 2x + y + 4z \\ y + z \end{pmatrix}$$

$$U: \begin{pmatrix} x \\ y \\ z \end{pmatrix} \mapsto \begin{pmatrix} x + 3y - 2z \\ -2x - 9y + 5z \\ x + 10y + 4z \end{pmatrix}$$

find the matrix that represents (a) T (b) U (c) TU (d) UT.

8 Given that

$$T: \begin{pmatrix} x \\ y \\ z \end{pmatrix} \mapsto \begin{pmatrix} x - y + 3z \\ 2x + y + 4z \\ y + z \end{pmatrix}$$

$$U: \begin{pmatrix} x \\ y \\ z \end{pmatrix} \mapsto \begin{pmatrix} x + 3y - 2z \\ -2x - 9y + 5z \\ x + 10y + 4z \end{pmatrix}$$

find the matrix that represents
(a) T^{-1} (b) U^{-1} (c) $(TU)^{-1}$ (d) $(UT)^{-1}$.

9 Given that

$$\mathbf{A} = \begin{pmatrix} 1 & 0 & 2 \\ 4 & 1 & 7 \\ -2 & 3 & 6 \end{pmatrix} \quad \text{and} \quad \mathbf{B} = \begin{pmatrix} -2 & 1 & 3 \\ 4 & -9 & 5 \\ 1 & 1 & -2 \end{pmatrix}$$

find (a) $\mathbf{A}^{\mathrm{T}}$ (b) $\mathbf{B}^{\mathrm{T}}$.
Hence find (c) $(\mathbf{AB})^{\mathrm{T}}$ (d) $(\mathbf{BA})^{\mathrm{T}}$.

10 Given that

$$\mathbf{A} = \begin{pmatrix} 2 & -1 & 3 \\ 1 & 4 & -7 \\ 6 & 6 & -3 \end{pmatrix} \quad \text{and} \quad \mathbf{B} = \begin{pmatrix} 1 & 2 & 4 \\ 3 & 1 & -2 \\ -1 & 1 & 3 \end{pmatrix}$$

find (a) $\mathbf{A}^{\mathrm{T}}$ (b) $\mathbf{B}^{\mathrm{T}}$.
Hence find (c) $(\mathbf{AB})^{\mathrm{T}}$ (d) $(\mathbf{BA})^{\mathrm{T}}$.

3.10 Eigenvalues and eigenvectors

If you have a matrix

$$\begin{pmatrix} 4 & 1 \\ 2 & 3 \end{pmatrix}$$

and use it to transform the vector $\begin{pmatrix} 1 \\ 1 \end{pmatrix}$ you get

$$\begin{pmatrix} 4 & 1 \\ 2 & 3 \end{pmatrix} \begin{pmatrix} 1 \\ 1 \end{pmatrix} = \begin{pmatrix} 5 \\ 5 \end{pmatrix} = 5 \begin{pmatrix} 1 \\ 1 \end{pmatrix}$$

That is, the result is a multiple of the original vector.
Similarly:

$$\begin{pmatrix} 4 & 1 \\ 2 & 3 \end{pmatrix} \begin{pmatrix} 1 \\ -2 \end{pmatrix} = \begin{pmatrix} 2 \\ -4 \end{pmatrix} = 2 \begin{pmatrix} 1 \\ -2 \end{pmatrix}$$

Although this happens for the vectors $\begin{pmatrix} 1 \\ 1 \end{pmatrix}$ and $\begin{pmatrix} 1 \\ -2 \end{pmatrix}$, it is not *generally* true that the image vector is a multiple of the original vector. If you calculate $\begin{pmatrix} 4 & 1 \\ 2 & 3 \end{pmatrix} \begin{pmatrix} 3 \\ 5 \end{pmatrix}$ you obtain $\begin{pmatrix} 17 \\ 21 \end{pmatrix}$, which is *not* a multiple of $\begin{pmatrix} 3 \\ 5 \end{pmatrix}$.

■ **If A is a matrix and v is a non-zero vector such that $\mathbf{Av} = \lambda \mathbf{v}$ where λ is a scalar, then v is called an eigenvector of A and λ is called an eigenvalue.**

Now consider an equation

$$\mathbf{Ax} = \mathbf{0}$$

where **A** is a matrix and **x** is a non-zero column vector. Assume that **A** is non-singular. Then **A** has an inverse, $\mathbf{A}^{-1}$, and you have

$$\mathbf{A}^{-1}(\mathbf{Ax}) = \mathbf{A}^{-1}\mathbf{0} = \mathbf{0}$$

But $$\mathbf{A}^{-1}(\mathbf{Ax}) = (\mathbf{A}^{-1}\mathbf{A})\mathbf{x} = \mathbf{Ix} = \mathbf{x}$$

If $$\mathbf{A}^{-1}(\mathbf{Ax}) = \mathbf{0}$$

and also $$\mathbf{A}^{-1}(\mathbf{Ax}) = \mathbf{x}$$

then $$\mathbf{x} = \mathbf{0}$$

But it is stated above that $\mathbf{x} \neq \mathbf{0}$. So if $\mathbf{Ax} = \mathbf{0}$ and $\mathbf{x} \neq \mathbf{0}$ then the assumption is false and **A** *must be* singular: it doesn't have an inverse.

That is: $$\det \mathbf{A} = 0$$

If you now go back to the definition of an eigenvector you have $\mathbf{Av} = \lambda\mathbf{v}$, where $\mathbf{v} \neq \mathbf{0}$.

That is: $$\mathbf{Av} - \lambda\mathbf{v} = \mathbf{0}$$

or $$(\mathbf{A} - \lambda\mathbf{I})\mathbf{v} = \mathbf{0}$$

But from what you have just seen, if

$$(\mathbf{A} - \lambda\mathbf{I})\mathbf{v} = \mathbf{0} \quad \text{and} \quad \mathbf{v} \neq \mathbf{0}$$

then $$|\mathbf{A} - \lambda\mathbf{I}| = \mathbf{0}$$

■ $|\mathbf{A} - \lambda\,\mathbf{I}| = 0$ **is called the characteristic equation of A.**

If you solve the characteristic equation you can find the values of λ which are the eigenvalues of **A**.

Example 12

Find the eigenvalues and the corresponding eigenvectors of the matrix $\mathbf{A} = \begin{pmatrix} -2 & -2 \\ 1 & -5 \end{pmatrix}$.

The characteristic equation of **A** is $|\mathbf{A} - \lambda\mathbf{I}| = 0$, where λ is an eigenvalue of **A**.

That is: $$\left| \begin{pmatrix} -2 & -2 \\ 1 & -5 \end{pmatrix} - \begin{pmatrix} \lambda & 0 \\ 0 & \lambda \end{pmatrix} \right| = 0$$

or $$\begin{vmatrix} -2 - \lambda & -2 \\ 1 & -5 - \lambda \end{vmatrix} = 0$$

$$(-2 - \lambda)(-5 - \lambda) - (-2) = 0$$

$$10 + 2\lambda + 5\lambda + \lambda^2 + 2 = 0$$

$$\lambda^2 + 7\lambda + 12 = 0$$

$$(\lambda + 3)(\lambda + 4) = 0$$

$$\lambda = -3, \ -4$$

So $\lambda = -3, -4$ are the eigenvalues. Now if -3 is an eigenvalue then

$$\begin{pmatrix} -2 & -2 \\ 1 & -5 \end{pmatrix} \begin{pmatrix} x \\ y \end{pmatrix} = -3 \begin{pmatrix} x \\ y \end{pmatrix}$$

that is:

$$\begin{pmatrix} -2x - 2y \\ x - 5y \end{pmatrix} = \begin{pmatrix} -3x \\ -3y \end{pmatrix}$$

So:

$$-2x - 2y = -3x$$
$$\Rightarrow \quad -2y = -x$$
$$\Rightarrow \quad x = 2y$$

Notice that you obtain the same relationship from $x - 5y = -3y$. So if $\begin{pmatrix} x \\ y \end{pmatrix}$ is an eigenvector then $x = 2y$. Consequently the eigenvector can be written:

$$\begin{pmatrix} 2y \\ y \end{pmatrix} = y \begin{pmatrix} 2 \\ 1 \end{pmatrix}$$

so $\begin{pmatrix} 2 \\ 1 \end{pmatrix}$ or any multiple of $\begin{pmatrix} 2 \\ 1 \end{pmatrix}$ is an eigenvector. Under these circumstances you take $\begin{pmatrix} 2 \\ 1 \end{pmatrix}$ to be the eigenvector.

If -4 is an eigenvalue then

$$\begin{pmatrix} -2 & -2 \\ 1 & -5 \end{pmatrix} \begin{pmatrix} x \\ y \end{pmatrix} = -4 \begin{pmatrix} x \\ y \end{pmatrix}$$

that is:

$$\begin{pmatrix} -2x - 2y \\ x - 5y \end{pmatrix} = \begin{pmatrix} -4x \\ -4y \end{pmatrix}$$

So:

$$-2x - 2y = -4x$$
$$\Rightarrow \quad -2y = -2x$$
$$\Rightarrow \quad x = y$$

(Again, you can check this by ensuring that the second equation $x - 5y = -4y$ gives the same relationship.)

So an eigenvector $\begin{pmatrix} x \\ y \end{pmatrix}$ with $x = y$ is of the form $\begin{pmatrix} x \\ x \end{pmatrix} = x \begin{pmatrix} 1 \\ 1 \end{pmatrix}$.

So $\begin{pmatrix} 1 \\ 1 \end{pmatrix}$ is an eigenvector.

Example 13

Find the eigenvectors and corresponding eigenvalues of

$$\begin{pmatrix} 2 & -1 & 1 \\ 0 & 2 & 0 \\ 1 & 3 & 2 \end{pmatrix}$$

If $\begin{pmatrix} x \\ y \\ z \end{pmatrix}$ is an eigenvector and λ the corresponding eigenvalue then:

$$\begin{pmatrix} 2 & -1 & 1 \\ 0 & 2 & 0 \\ 1 & 3 & 2 \end{pmatrix} \begin{pmatrix} x \\ y \\ z \end{pmatrix} = \lambda \begin{pmatrix} x \\ y \\ z \end{pmatrix}$$

So the characteristic equation is

$$\left| \begin{pmatrix} 2 & -1 & 1 \\ 0 & 2 & 0 \\ 1 & 3 & 2 \end{pmatrix} - \lambda \begin{pmatrix} 1 & 0 & 0 \\ 0 & 1 & 0 \\ 0 & 0 & 1 \end{pmatrix} \right| = 0$$

$$\Rightarrow \quad \begin{vmatrix} 2-\lambda & -1 & 1 \\ 0 & 2-\lambda & 0 \\ 1 & 3 & 2-\lambda \end{vmatrix} = 0$$

Expanding the determinant gives

$$(2-\lambda)(2-\lambda)(2-\lambda) + 1(-1)(2-\lambda) = 0$$

$$\Rightarrow \quad (2-\lambda)\left[(2-\lambda)^2 - 1\right] = 0$$

$$(2-\lambda)(2-\lambda-1)(2-\lambda+1) = 0$$

using the difference of two squares.

$\lambda = 2, 1, 3$

For $\lambda = 1$:

$$\begin{pmatrix} 2 & -1 & 1 \\ 0 & 2 & 0 \\ 1 & 3 & 2 \end{pmatrix} \begin{pmatrix} x \\ y \\ z \end{pmatrix} = 1 \begin{pmatrix} x \\ y \\ z \end{pmatrix}$$

$$\begin{pmatrix} 2x - y + z \\ 2y \\ x + 3y + 2z \end{pmatrix} = \begin{pmatrix} x \\ y \\ z \end{pmatrix}$$

From the second equation:

$$2y = y \Rightarrow y = 0$$

From the first equation:

$$2x + z = x \Rightarrow z = -x$$

The eigenvector is of the form $\begin{pmatrix} x \\ 0 \\ -x \end{pmatrix}$.

So an eigenvector corresponding to 1 is $\begin{pmatrix} 1 \\ 0 \\ -1 \end{pmatrix}$.

For $\lambda = 2$

$$\begin{pmatrix} 2 & -1 & 1 \\ 0 & 2 & 0 \\ 1 & 3 & 2 \end{pmatrix} \begin{pmatrix} x \\ y \\ z \end{pmatrix} = 2 \begin{pmatrix} x \\ y \\ z \end{pmatrix}$$

$$\begin{pmatrix} 2x - y + z \\ 2y \\ x + 3y + 2z \end{pmatrix} = \begin{pmatrix} 2x \\ 2y \\ 2z \end{pmatrix}$$

$$2x - y + z = 2x \implies y = z$$

$$x + 3y + 2z = 2z \implies x + 3y = 0$$

So
$$x = -3y$$

An eigenvector is $\begin{pmatrix} -3y \\ y \\ y \end{pmatrix} = y \begin{pmatrix} -3 \\ 1 \\ 1 \end{pmatrix}$.

So an eigenvector corresponding to 2 is $\begin{pmatrix} -3 \\ 1 \\ 1 \end{pmatrix}$.

Finally, if $\lambda = 3$ then

$$\begin{pmatrix} 2x - y + z \\ 2y \\ x + 3y + 2z \end{pmatrix} = \begin{pmatrix} 3x \\ 3y \\ 3z \end{pmatrix}$$

$$2y = 3y \implies y = 0$$

$$2x + z = 3x \implies x = z$$

So an eigenvector is $\begin{pmatrix} x \\ 0 \\ x \end{pmatrix} = x \begin{pmatrix} 1 \\ 0 \\ 1 \end{pmatrix}$.

An eigenvector corresponding to 3 is $\begin{pmatrix} 1 \\ 0 \\ 1 \end{pmatrix}$.

Example 14

Find the eigenvalues and corresponding eigenvectors of

$$\begin{pmatrix} 3 & 2 & 4 \\ 2 & 0 & 2 \\ 4 & 2 & 3 \end{pmatrix}$$

The characteristic equation is

$$\begin{vmatrix} 3-\lambda & 2 & 4 \\ 2 & -\lambda & 2 \\ 4 & 2 & 3-\lambda \end{vmatrix} = 0$$

That is:

$$(3-\lambda)(-3\lambda+\lambda^2-4) - 2(6-2\lambda-8) + 4(4+4\lambda) = 0$$
$$(3-\lambda)(\lambda^2-3\lambda-4) - 2(-2\lambda-2) + 16(\lambda+1) = 0$$
$$(3-\lambda)(\lambda+1)(\lambda-4) + 4(\lambda+1) + 16(\lambda+1) = 0$$
$$(\lambda+1)(3\lambda-12-\lambda^2+4\lambda+4+16) = 0$$
$$(\lambda+1)(-\lambda^2+7\lambda+8) = 0$$
$$(\lambda+1)(8-\lambda)(1+\lambda) = 0$$
$$\lambda = -1, \ -1, 8$$

For $\lambda = 8$:

$$\begin{pmatrix} 3 & 2 & 4 \\ 2 & 0 & 2 \\ 4 & 2 & 3 \end{pmatrix}\begin{pmatrix} x \\ y \\ z \end{pmatrix} = 8\begin{pmatrix} x \\ y \\ z \end{pmatrix}$$

$$\begin{pmatrix} 3x+2y+4z \\ 2x+2z \\ 4x+2y+3z \end{pmatrix} = \begin{pmatrix} 8x \\ 8y \\ 8z \end{pmatrix}$$

$$3x+2y+4z = 8x$$
$$\Rightarrow \quad -5x+2y+4z = 0 \qquad\qquad (1)$$

$$2x+2z = 8y$$
$$\Rightarrow \quad x-4y+z = 0 \qquad\qquad (2)$$

(1) × 2 gives $\qquad -10x+4y+8z = 0 \qquad\qquad (3)$

(2) + (3) gives $\qquad -9x+9z = 0$

So: $\qquad\qquad\qquad x = z$

Substituting in (1) gives

$$-5z+2y+4z = 0$$

$$2y = z$$

or $\qquad\qquad\qquad y = \tfrac{1}{2}z$

So an eigenvector is $\begin{pmatrix} z \\ \frac{1}{2}z \\ z \end{pmatrix}$ or $z\begin{pmatrix} 1 \\ \frac{1}{2} \\ 1 \end{pmatrix}$.

An eigenvector corresponding to 8 is $\begin{pmatrix} 2 \\ 1 \\ 2 \end{pmatrix}$.

For $\lambda = -1$:

$$\begin{pmatrix} 3x + 2y + 4z \\ 2x + 2z \\ 4x + 2y + 3z \end{pmatrix} = \begin{pmatrix} -x \\ -y \\ -z \end{pmatrix}$$

$$3x + 2y + 4z = -x$$

$$\Rightarrow \quad 2x + y + 2z = 0 \tag{1}$$

$$2x + 2z = -y$$

$$\Rightarrow \quad 2x + y + 2z = 0 \tag{2}$$

$$4x + 2y + 3z = -z$$

$$\Rightarrow \quad 2x + y + 2z = 0 \tag{3}$$

Now (1), (2) and (3) are identical so you cannot solve them to find a relationship between x, y and z. This has happened because -1 is a repeated eigenvalue.

Under these circumstances you can choose any two independent vectors which satisfy $2x + y + 2z = 0$. 'Independent' means that one must not be a multiple of the other.

So, in this case,

$$\begin{pmatrix} 1 \\ 0 \\ -1 \end{pmatrix} \quad \text{and} \quad \begin{pmatrix} 1 \\ -2 \\ 0 \end{pmatrix}$$

are appropriate because they are independent and

$$(2 \times 1) + 0 + 2(-1) = 2 - 2 = 0$$

and

$$(2 \times 1) + (-2) + (2 \times 0) = 2 - 2 = 0$$

So the eigenvalues are -1, -1 and 8 with corresponding eigenvectors

$$\begin{pmatrix} 1 \\ 0 \\ -1 \end{pmatrix}, \begin{pmatrix} 1 \\ -2 \\ 0 \end{pmatrix} \quad \text{and} \quad \begin{pmatrix} 2 \\ 1 \\ 2 \end{pmatrix}.$$

Normalised eigenvectors

If you have a vector $x\mathbf{i} + y\mathbf{j} + z\mathbf{k}$ then you can make this into a unit vector by dividing by $\sqrt{(x^2 + y^2 + z^2)}$ (Book C4, page 51). That is,

$$\frac{x}{\sqrt{(x^2 + y^2 + z^2)}}\mathbf{i} + \frac{y}{\sqrt{(x^2 + y^2 + z^2)}}\mathbf{j} + \frac{z}{\sqrt{(x^2 + y^2 + z^2)}}\mathbf{k}$$

is a unit vector.

In the same way you can transform an eigenvector into a unit eigenvector. If $\begin{pmatrix} x \\ y \\ z \end{pmatrix}$ is an eigenvector then

$$\begin{pmatrix} \dfrac{x}{\sqrt{(x^2 + y^2 + z^2)}} \\ \dfrac{y}{\sqrt{(x^2 + y^2 + z^2)}} \\ \dfrac{z}{\sqrt{(x^2 + y^2 + z^2)}} \end{pmatrix}$$

is an eigenvector with unit length. Such eigenvectors are not usually *called* unit eigenvectors (although this is what they are). They are usually called **normalised eigenvectors**.

■ **If** $\begin{pmatrix} x \\ y \\ z \end{pmatrix}$ **is an eigenvector then**

$$\begin{pmatrix} \dfrac{x}{\sqrt{(x^2 + y^2 + z^2)}} \\ \dfrac{y}{\sqrt{(x^2 + y^2 + z^2)}} \\ \dfrac{z}{\sqrt{(x^2 + y^2 + z^2)}} \end{pmatrix}$$

is the corresponding normalised eigenvector.

Example 15

Normalise the eigenvector $\begin{pmatrix} 1 \\ 4 \\ -6 \end{pmatrix}$.

$$\sqrt{(1^2 + 4^2 + (-6)^2)} = \sqrt{(1 + 16 + 36)}$$
$$= \sqrt{53}$$

so the corresponding normalised eigenvector is

$$\begin{pmatrix} \frac{1}{\sqrt{53}} \\ \frac{4}{\sqrt{53}} \\ -\frac{6}{\sqrt{53}} \end{pmatrix}$$

Orthogonal eigenvectors

If two vectors $\mathbf{x}_1$ and $\mathbf{x}_2$ are perpendicular, then the scalar product $\mathbf{x}_1.\mathbf{x}_2 = 0$ (Book P3, page 163). The same is true for eigenvectors.

If the eigenvector $\mathbf{x}_1 = \begin{pmatrix} 2 \\ -1 \\ 3 \end{pmatrix}$ and the eigenvector $\mathbf{x}_2 = \begin{pmatrix} 2 \\ 7 \\ 1 \end{pmatrix}$

then their scalar product is

$$\mathbf{x}_1.\mathbf{x}_2 = \begin{pmatrix} 2 \\ -1 \\ 3 \end{pmatrix}.\begin{pmatrix} 2 \\ 7 \\ 1 \end{pmatrix}$$

$$= (2 \times 2) + (-1 \times 7) + (3 \times 1)$$

$$= 4 - 7 + 3 = 0$$

So the eigenvectors $\begin{pmatrix} 2 \\ -1 \\ 3 \end{pmatrix}$ and $\begin{pmatrix} 2 \\ 7 \\ 1 \end{pmatrix}$ are perpendicular to each

other. In this context the term 'perpendicular' is not usually used. Instead, it is more common to say that the eigenvectors are **orthogonal**.

■ **If two eigenvectors**

$$\begin{pmatrix} a_1 \\ a_2 \\ a_3 \end{pmatrix} \text{ and } \begin{pmatrix} b_1 \\ b_2 \\ b_3 \end{pmatrix}$$

are such that their scalar product is zero, then the eigenvectors are said to be orthogonal.

Example 16

Find a vector which is orthogonal to $\begin{pmatrix} 1 \\ 2 \\ -1 \end{pmatrix}$.

If $\begin{pmatrix} x \\ y \\ z \end{pmatrix}$ is orthogonal to $\begin{pmatrix} 1 \\ 2 \\ -1 \end{pmatrix}$ then

$$\begin{pmatrix} x \\ y \\ z \end{pmatrix} \cdot \begin{pmatrix} 1 \\ 2 \\ -1 \end{pmatrix} = 0$$

That is:
$$x + 2y - z = 0$$

There are many values of x, y and z which will satisfy this condition. One such set is $x = 1$, $y = 3$, $z = 7$ because

$$1 + (2 \times 3) - 7 = 0$$

So $\begin{pmatrix} 1 \\ 3 \\ 7 \end{pmatrix}$ is orthogonal to $\begin{pmatrix} 1 \\ 2 \\ -1 \end{pmatrix}$.

Orthogonal matrices

You can see that the vectors

$$\begin{pmatrix} 1 \\ 0 \\ -1 \end{pmatrix}, \begin{pmatrix} -1 \\ 1 \\ -1 \end{pmatrix}, \begin{pmatrix} 1 \\ 2 \\ 1 \end{pmatrix}$$

are orthogonal to each other, because

$$\begin{pmatrix} 1 \\ 0 \\ -1 \end{pmatrix} \cdot \begin{pmatrix} -1 \\ 1 \\ -1 \end{pmatrix} = -1 + 0 + 1 = 0$$

$$\begin{pmatrix} 1 \\ 0 \\ -1 \end{pmatrix} \cdot \begin{pmatrix} 1 \\ 2 \\ 1 \end{pmatrix} = 1 + 0 - 1 = 0$$

and
$$\begin{pmatrix} -1 \\ 1 \\ -1 \end{pmatrix} \cdot \begin{pmatrix} 1 \\ 2 \\ 1 \end{pmatrix} = -1 + 2 - 1 = 0$$

If you now normalise each of the vectors they become

$$\begin{pmatrix} \frac{1}{\sqrt{2}} \\ 0 \\ -\frac{1}{\sqrt{2}} \end{pmatrix}, \begin{pmatrix} -\frac{1}{\sqrt{3}} \\ \frac{1}{\sqrt{3}} \\ -\frac{1}{\sqrt{3}} \end{pmatrix} \quad \text{and} \quad \begin{pmatrix} \frac{1}{\sqrt{6}} \\ \frac{2}{\sqrt{6}} \\ \frac{1}{\sqrt{6}} \end{pmatrix}$$

The matrix
$$\begin{pmatrix} \frac{1}{\sqrt{2}} & -\frac{1}{\sqrt{3}} & \frac{1}{\sqrt{6}} \\ 0 & \frac{1}{\sqrt{3}} & \frac{2}{\sqrt{6}} \\ -\frac{1}{\sqrt{2}} & -\frac{1}{\sqrt{3}} & \frac{1}{\sqrt{6}} \end{pmatrix}$$

whose columns are those of normalised vectors *which are also* orthogonal to each other is called an **orthogonal matrix**.

■ If $\mathbf{M} = \begin{pmatrix} a_1 & b_1 & c_1 \\ a_2 & b_2 & c_2 \\ a_3 & b_3 & c_3 \end{pmatrix}$

where $\begin{pmatrix} a_1 \\ a_2 \\ a_3 \end{pmatrix}$, $\begin{pmatrix} b_1 \\ b_2 \\ b_3 \end{pmatrix}$ and $\begin{pmatrix} c_1 \\ c_2 \\ c_3 \end{pmatrix}$ are each **normalised vectors**

and if $\begin{pmatrix} a_1 \\ a_2 \\ a_3 \end{pmatrix} \cdot \begin{pmatrix} b_1 \\ b_2 \\ b_3 \end{pmatrix} = 0$, $\begin{pmatrix} a_1 \\ a_2 \\ a_3 \end{pmatrix} \cdot \begin{pmatrix} c_1 \\ c_2 \\ c_3 \end{pmatrix} = 0$

and $\begin{pmatrix} b_1 \\ b_2 \\ b_3 \end{pmatrix} \cdot \begin{pmatrix} c_1 \\ c_2 \\ c_3 \end{pmatrix} = 0$, then M is called an **orthogonal matrix**.

Now, you have seen above that $\mathbf{M} = \begin{pmatrix} \frac{1}{\sqrt{2}} & -\frac{1}{\sqrt{3}} & \frac{1}{\sqrt{6}} \\ 0 & \frac{1}{\sqrt{3}} & \frac{2}{\sqrt{6}} \\ -\frac{1}{\sqrt{2}} & -\frac{1}{\sqrt{3}} & \frac{1}{\sqrt{6}} \end{pmatrix}$

is an orthogonal matrix.

Then $\mathbf{M}^{\mathrm{T}} = \begin{pmatrix} \frac{1}{\sqrt{2}} & 0 & -\frac{1}{\sqrt{2}} \\ -\frac{1}{\sqrt{3}} & \frac{1}{\sqrt{3}} & -\frac{1}{\sqrt{3}} \\ \frac{1}{\sqrt{6}} & \frac{2}{\sqrt{6}} & \frac{1}{\sqrt{6}} \end{pmatrix}$

So: $\mathbf{M}^{\mathrm{T}}\mathbf{M} = \begin{pmatrix} \frac{1}{\sqrt{2}} & 0 & -\frac{1}{\sqrt{2}} \\ -\frac{1}{\sqrt{3}} & \frac{1}{\sqrt{3}} & -\frac{1}{\sqrt{3}} \\ \frac{1}{\sqrt{6}} & \frac{2}{\sqrt{6}} & \frac{1}{\sqrt{6}} \end{pmatrix} \begin{pmatrix} \frac{1}{\sqrt{2}} & -\frac{1}{\sqrt{3}} & \frac{1}{\sqrt{6}} \\ 0 & \frac{1}{\sqrt{3}} & \frac{2}{\sqrt{6}} \\ -\frac{1}{\sqrt{2}} & -\frac{1}{\sqrt{3}} & \frac{1}{\sqrt{6}} \end{pmatrix}$

$= \begin{pmatrix} 1 & 0 & 0 \\ 0 & 1 & 0 \\ 0 & 0 & 1 \end{pmatrix}$

Hence, in this case, $\mathbf{M}^{\mathrm{T}} = \mathbf{M}^{-1}$.
In fact, this is true for all orthogonal matrices.

If $\mathbf{a} = \begin{pmatrix} a_1 \\ a_2 \\ a_3 \end{pmatrix}$, $\mathbf{b} = \begin{pmatrix} b_1 \\ b_2 \\ b_3 \end{pmatrix}$ and $\mathbf{c} = \begin{pmatrix} c_1 \\ c_2 \\ c_3 \end{pmatrix}$ are normalised vectors

such that $\mathbf{a}.\mathbf{b} = \mathbf{b}.\mathbf{c} = \mathbf{c}.\mathbf{a} = 0$, then $\mathbf{M} = \begin{pmatrix} a_1 & b_1 & c_1 \\ a_2 & b_2 & c_2 \\ a_3 & b_3 & c_3 \end{pmatrix}$ is

orthogonal.

So: $\qquad\qquad \mathbf{M}^{\mathrm{T}} = \begin{pmatrix} a_1 & a_2 & a_3 \\ b_1 & b_2 & b_3 \\ c_1 & c_2 & c_3 \end{pmatrix}$

Then: $\qquad \mathbf{M}^{\mathrm{T}}\mathbf{M} = \begin{pmatrix} a_1 & a_2 & a_3 \\ b_1 & b_2 & b_3 \\ c_1 & c_2 & c_3 \end{pmatrix} \begin{pmatrix} a_1 & b_1 & c_1 \\ a_2 & b_2 & c_2 \\ a_3 & b_3 & c_3 \end{pmatrix}$

$$= \begin{pmatrix} 1 & 0 & 0 \\ 0 & 1 & 0 \\ 0 & 0 & 1 \end{pmatrix}$$

since $\qquad\qquad \begin{pmatrix} a_1 \\ a_2 \\ a_3 \end{pmatrix} . \begin{pmatrix} a_1 \\ a_2 \\ a_3 \end{pmatrix} = 1$

because $\mathbf{a}$ has been normalised,

$$\begin{pmatrix} b_1 \\ b_2 \\ b_3 \end{pmatrix} . \begin{pmatrix} b_1 \\ b_2 \\ b_3 \end{pmatrix} = 1$$

because $\mathbf{b}$ has been normalised and

$$\begin{pmatrix} c_1 \\ c_2 \\ c_3 \end{pmatrix} . \begin{pmatrix} c_1 \\ c_2 \\ c_3 \end{pmatrix} = 1$$

because $\mathbf{c}$ has been normalised and

$$\begin{pmatrix} a_1 \\ a_2 \\ a_3 \end{pmatrix} . \begin{pmatrix} b_1 \\ b_2 \\ b_3 \end{pmatrix} = \begin{pmatrix} a_1 \\ a_2 \\ a_3 \end{pmatrix} . \begin{pmatrix} c_1 \\ c_2 \\ c_3 \end{pmatrix} = \begin{pmatrix} b_1 \\ b_2 \\ b_3 \end{pmatrix} . \begin{pmatrix} c_1 \\ c_2 \\ c_3 \end{pmatrix} = 0$$

because $\mathbf{a}$, $\mathbf{b}$ and $\mathbf{c}$ are mutually orthogonal.

- **For any orthogonal matrix M,**

$$\mathbf{M}^{\mathrm{T}} = \mathbf{M}^{-1}$$

3.11 Diagonalising a symmetric matrix

Square matrices of the form

$$\begin{pmatrix} a & 0 \\ 0 & b \end{pmatrix} \quad \text{and} \quad \begin{pmatrix} a & 0 & 0 \\ 0 & b & 0 \\ 0 & 0 & c \end{pmatrix}$$

which have numbers in the diagonal that goes from the top left-hand corner to the bottom right-hand corner, and zeros everywhere else, are called **diagonal matrices**.

So $\begin{pmatrix} 1 & 0 \\ 0 & -3 \end{pmatrix}$ and $\begin{pmatrix} 2 & 0 & 0 \\ 0 & 1 & 0 \\ 0 & 0 & 7 \end{pmatrix}$ are diagonal matrices.

Any square matrix $\mathbf{M}$ which is such that $\mathbf{M}^{\mathrm{T}} = \mathbf{M}$ is called a **symmetric matrix**. So $\mathbf{A} = \begin{pmatrix} 2 & 5 \\ 5 & -1 \end{pmatrix}$ is symmetric because

$$\mathbf{A}^{\mathrm{T}} = \begin{pmatrix} 2 & 5 \\ 5 & -1 \end{pmatrix} = \mathbf{A}$$

Similarly

$$\mathbf{B} = \begin{pmatrix} 2 & -3 & 1 \\ -3 & 5 & 24 \\ 1 & 24 & 10 \end{pmatrix}$$

is symmetric because

$$\mathbf{B}^{\mathrm{T}} = \begin{pmatrix} 2 & -3 & 1 \\ -3 & 5 & 24 \\ 1 & 24 & 10 \end{pmatrix} = \mathbf{B}$$

Notice that each of these matrices is symmetrical about its *leading* diagonal. That is, the diagonal from top left to bottom right.

Now for some purposes in higher mathematics it is necessary to diagonalise a matrix: that is, to transform the matrix to a diagonal matrix. This is not always possible to do. However, if the original matrix is a real symmetric 3×3 matrix then the matrix always has three real eigenvalues (some of these may be repeated eigenvalues) and you can find three corresponding eigenvectors which are mutually orthogonal. As a result, you can always diagonalise a real symmetric 3×3 matrix.

The next example shows you how to do this.

Example 16

Find the eigenvalues and corresponding orthogonal eigenvectors of the matrix

$$\mathbf{A} = \begin{pmatrix} 2 & 0 & -2 \\ 0 & 4 & 0 \\ -2 & 0 & 5 \end{pmatrix}$$

Hence find a matrix $\mathbf{P}$ such that $\mathbf{P}^{\mathrm{T}}\mathbf{A}\mathbf{P}$ is a diagonal matrix.

Because $\mathbf{A}$ is symmetric it has three real eigenvalues and it is possible to find three corresponding, mutually orthogonal eigenvectors.

The characteristic equation is

$$\begin{vmatrix} 2-\lambda & 0 & -2 \\ 0 & 4-\lambda & 0 \\ -2 & 0 & 5-\lambda \end{vmatrix} = 0$$

Expanding the determinant gives

$$(2-\lambda)(4-\lambda)(5-\lambda) - 2 \times 2(4-\lambda) = 0$$

$$(4-\lambda)[(2-\lambda)(5-\lambda) - 4] = 0$$

$$(4-\lambda)(\lambda^2 - 7\lambda + 10 - 4) = 0$$

$$(4-\lambda)(\lambda^2 - 7\lambda + 6) = 0$$

$$(4-\lambda)(\lambda - 6)(\lambda - 1) = 0$$

So: $\qquad\qquad\qquad \lambda = 1, 4, 6$

For $\lambda = 1$:
$$\begin{pmatrix} 2 & 0 & -2 \\ 0 & 4 & 0 \\ -2 & 0 & 5 \end{pmatrix}\begin{pmatrix} x \\ y \\ z \end{pmatrix} = 1\begin{pmatrix} x \\ y \\ z \end{pmatrix}$$

That is:
$$\begin{pmatrix} 2x - 2z \\ 4y \\ -2x + 5z \end{pmatrix} = \begin{pmatrix} x \\ y \\ z \end{pmatrix}$$

From the second line: $\qquad\qquad 4y = y \Rightarrow y = 0$

From the first line: $\qquad\qquad 2x - 2z = x \Rightarrow x = 2z$

So an eigenvector corresponding to $\lambda = 1$ is $\begin{pmatrix} 2 \\ 0 \\ 1 \end{pmatrix}$

Normalised this is: $\begin{pmatrix} \frac{2}{\sqrt{5}} \\ 0 \\ \frac{1}{\sqrt{5}} \end{pmatrix}$

For $\lambda = 4$:

$$\begin{pmatrix} 2x - 2z \\ 4y \\ -2x + 5z \end{pmatrix} = \begin{pmatrix} 4x \\ 4y \\ 4z \end{pmatrix}$$

From the first line: $\qquad\qquad\qquad\qquad\qquad x = -z$

From the third line: $\qquad\qquad\qquad\qquad\qquad z = 2x$

These are inconsistent, so the only values of x and z which will satisfy them both are $x = z = 0$.

An eigenvector corresponding to $\lambda = 4$ is $\begin{pmatrix} 0 \\ 1 \\ 0 \end{pmatrix}$, which is already normalised.

For $\lambda = 6$:

$$\begin{pmatrix} 2x - 2z \\ 4y \\ -2x + 5z \end{pmatrix} = \begin{pmatrix} 6x \\ 6y \\ 6z \end{pmatrix}$$

From the first line: $\qquad\qquad\qquad -2z = 4x \Rightarrow z = -2x$

From the second line: $\qquad\qquad\qquad 4y = 6y \Rightarrow y = 0$

So an eigenvector corresponding to $\lambda = 6$ is

$\begin{pmatrix} 1 \\ 0 \\ -2 \end{pmatrix}$, which when normalised is $\begin{pmatrix} \frac{1}{\sqrt{5}} \\ 0 \\ -\frac{2}{\sqrt{5}} \end{pmatrix}$.

Now $\qquad\qquad \begin{pmatrix} \frac{2}{\sqrt{5}} \\ 0 \\ \frac{1}{\sqrt{5}} \end{pmatrix} \cdot \begin{pmatrix} 0 \\ 1 \\ 0 \end{pmatrix} = 0$

$$\begin{pmatrix} \frac{2}{\sqrt{5}} \\ 0 \\ \frac{1}{\sqrt{5}} \end{pmatrix} \cdot \begin{pmatrix} \frac{1}{\sqrt{5}} \\ 0 \\ -\frac{2}{\sqrt{5}} \end{pmatrix} = \frac{2}{5} + 0 - \frac{2}{5} = 0$$

and $\qquad\qquad \begin{pmatrix} 0 \\ 1 \\ 0 \end{pmatrix} \cdot \begin{pmatrix} \frac{1}{\sqrt{5}} \\ 0 \\ -\frac{2}{\sqrt{5}} \end{pmatrix} = 0$

So the normalised eigenvectors are also orthogonal.

Choose **P** to be the matrix whose columns are the normalised, orthogonal eigenvectors:

$$\mathbf{P} = \begin{pmatrix} \frac{2}{\sqrt{5}} & 0 & \frac{1}{\sqrt{5}} \\ 0 & 1 & 0 \\ \frac{1}{\sqrt{5}} & 0 & -\frac{2}{\sqrt{5}} \end{pmatrix}$$

Then:

$$\mathbf{P}^{\mathrm{T}} = \begin{pmatrix} \frac{2}{\sqrt{5}} & 0 & \frac{1}{\sqrt{5}} \\ 0 & 1 & 0 \\ \frac{1}{\sqrt{5}} & 0 & -\frac{2}{\sqrt{5}} \end{pmatrix}$$

So $\mathbf{P}^{\mathrm{T}}\mathbf{A}\mathbf{P} = \begin{pmatrix} \frac{2}{\sqrt{5}} & 0 & \frac{1}{\sqrt{5}} \\ 0 & 1 & 0 \\ \frac{1}{\sqrt{5}} & 0 & -\frac{2}{\sqrt{5}} \end{pmatrix} \begin{pmatrix} 2 & 0 & -2 \\ 0 & 4 & 0 \\ -2 & 0 & 5 \end{pmatrix} \begin{pmatrix} \frac{2}{\sqrt{5}} & 0 & \frac{1}{\sqrt{5}} \\ 0 & 1 & 0 \\ \frac{1}{\sqrt{5}} & 0 & -\frac{2}{\sqrt{5}} \end{pmatrix}$

$$= \begin{pmatrix} \frac{2}{\sqrt{5}} & 0 & \frac{1}{\sqrt{5}} \\ 0 & 1 & 0 \\ \frac{1}{\sqrt{5}} & 0 & -\frac{2}{\sqrt{5}} \end{pmatrix} \begin{pmatrix} \frac{2}{\sqrt{5}} & 0 & \frac{6}{\sqrt{5}} \\ 0 & 4 & 0 \\ \frac{1}{\sqrt{5}} & 0 & -\frac{12}{\sqrt{5}} \end{pmatrix}$$

$$= \begin{pmatrix} 1 & 0 & 0 \\ 0 & 4 & 0 \\ 0 & 0 & 6 \end{pmatrix}$$

which is a diagonal matrix, as required. Notice that the numbers 1, 4, 6 along the leading diagonal are the eigenvalues corresponding

to $\begin{pmatrix} \frac{2}{\sqrt{5}} \\ 0 \\ \frac{1}{\sqrt{5}} \end{pmatrix}$, $\begin{pmatrix} 0 \\ 1 \\ 0 \end{pmatrix}$ and $\begin{pmatrix} \frac{1}{\sqrt{5}} \\ 0 \\ -\frac{2}{\sqrt{5}} \end{pmatrix}$.

You should now be able to see why this process is *always* going to diagonalise a real symmetric matrix. Suppose that the symmetric matrix is

$$\mathbf{A} = \begin{pmatrix} p & q & r \\ q & s & t \\ r & t & u \end{pmatrix}$$

and suppose that it has eigenvalues λ_1, λ_2, λ_3 with corresponding normalised orthogonal eigenvectors

$$\begin{pmatrix} a_1 \\ a_2 \\ a_3 \end{pmatrix}, \begin{pmatrix} b_1 \\ b_2 \\ b_3 \end{pmatrix} \text{ and } \begin{pmatrix} c_1 \\ c_2 \\ c_3 \end{pmatrix}$$

So:
$$\begin{pmatrix} p & q & r \\ q & s & t \\ r & t & u \end{pmatrix} \begin{pmatrix} a_1 \\ a_2 \\ a_3 \end{pmatrix} = \lambda_1 \begin{pmatrix} a_1 \\ a_2 \\ a_3 \end{pmatrix}$$

$$\begin{pmatrix} p & q & r \\ q & s & t \\ r & t & u \end{pmatrix} \begin{pmatrix} b_1 \\ b_2 \\ b_3 \end{pmatrix} = \lambda_2 \begin{pmatrix} b_1 \\ b_2 \\ b_3 \end{pmatrix}$$

and
$$\begin{pmatrix} p & q & r \\ q & s & t \\ r & t & u \end{pmatrix} \begin{pmatrix} c_1 \\ c_2 \\ c_3 \end{pmatrix} = \lambda_3 \begin{pmatrix} c_1 \\ c_2 \\ c_3 \end{pmatrix}$$

Now:
$$\mathbf{P} = \begin{pmatrix} a_1 & b_1 & c_1 \\ a_2 & b_2 & c_2 \\ a_3 & b_3 & c_3 \end{pmatrix}$$

is orthogonal, and

$$\begin{aligned}
\mathbf{AP} &= \begin{pmatrix} p & q & r \\ q & s & t \\ r & t & u \end{pmatrix} \begin{pmatrix} a_1 & b_1 & c_1 \\ a_2 & b_2 & c_2 \\ a_3 & b_3 & c_3 \end{pmatrix} \\
&= \begin{pmatrix} \lambda_1 a_1 & \lambda_2 b_1 & \lambda_3 c_1 \\ \lambda_1 a_2 & \lambda_2 b_2 & \lambda_3 c_2 \\ \lambda_1 a_3 & \lambda_2 b_3 & \lambda_3 c_3 \end{pmatrix} \\
&= \begin{pmatrix} a_1 & b_1 & c_1 \\ a_2 & b_2 & c_2 \\ a_3 & b_3 & c_3 \end{pmatrix} \begin{pmatrix} \lambda_1 & 0 & 0 \\ 0 & \lambda_2 & 0 \\ 0 & 0 & \lambda_3 \end{pmatrix}
\end{aligned}$$

$= \mathbf{PD}$, say, where

$$\mathbf{D} = \begin{pmatrix} \lambda_1 & 0 & 0 \\ 0 & \lambda_2 & 0 \\ 0 & 0 & \lambda_3 \end{pmatrix}$$

Now
$$\begin{aligned}
\mathbf{P}^T \mathbf{A P} &= \mathbf{P}^{-1} \mathbf{A P}, && \text{since } \mathbf{P} \text{ is orthogonal} \\
&= \mathbf{P}^{-1}(\mathbf{PD}), && \text{since } \mathbf{AP} = \mathbf{PD} \\
&= (\mathbf{P}^{-1}\mathbf{P})\mathbf{D} \\
&= \mathbf{I D} \\
&= \mathbf{D}
\end{aligned}$$

where $\mathbf{D}$ is the matrix whose leading diagonal consists of the eigenvalues of $\mathbf{A}$.

■ **To diagonalise a 3×3 real symmetric matrix A, you find the normalised, orthogonal eigenvectors of A and form P whose columns consist of these eigenvectors. Then $\mathbf{P}^{\mathrm{T}}\mathbf{AP}$ is a diagonal matrix whose elements in the leading diagonal are the eigenvalues of A.**

Example 17

Given the symmetric matrix

$$\mathbf{A} = \begin{pmatrix} 1 & 0 & 2 \\ 0 & 4 & 0 \\ 2 & 0 & 1 \end{pmatrix}$$

find an orthogonal matrix **P** such that $\mathbf{P}^{\mathrm{T}}\mathbf{AP}$ is a diagonal matrix. Evaluate $\mathbf{P}^{\mathrm{T}}\mathbf{AP}$.

The characteristic equation is

$$\begin{vmatrix} 1-\lambda & 0 & 2 \\ 0 & 4-\lambda & 0 \\ 2 & 0 & 1-\lambda \end{vmatrix} = 0$$

Expanding the determinant gives

$$(1-\lambda)(4-\lambda)(1-\lambda) - 2(2)(4-\lambda) = 0$$
$$(4-\lambda)\left[(1-\lambda)^2 - 4\right] = 0$$
$$(4-\lambda)(1 - 2\lambda + \lambda^2 - 4) = 0$$
$$(4-\lambda)(\lambda^2 - 2\lambda - 3) = 0$$
$$(4-\lambda)(\lambda - 3)(\lambda + 1) = 0$$

So $\lambda = -1, 3, 4$

When $\lambda = -1$:

$$\begin{pmatrix} 1 & 0 & 2 \\ 0 & 4 & 0 \\ 2 & 0 & 1 \end{pmatrix}\begin{pmatrix} x \\ y \\ z \end{pmatrix} = \begin{pmatrix} -x \\ -y \\ -z \end{pmatrix}$$
$$\begin{pmatrix} x+2z \\ 4y \\ 2x+z \end{pmatrix} = \begin{pmatrix} -x \\ -y \\ -z \end{pmatrix}$$
$$\Rightarrow \quad x = -z \quad \text{and} \quad y = 0$$

$\begin{pmatrix} 1 \\ 0 \\ -1 \end{pmatrix}$ is an eigenvector and $\begin{pmatrix} \frac{1}{\sqrt{2}} \\ 0 \\ -\frac{1}{\sqrt{2}} \end{pmatrix}$ is the normalised eigenvector.

When $\lambda = 3$:

$$\begin{pmatrix} x + 2z \\ 4y \\ 2x + z \end{pmatrix} = \begin{pmatrix} 3x \\ 3y \\ 3z \end{pmatrix}$$

$$\Rightarrow \quad x = z \quad \text{and} \quad y = 0$$

$\begin{pmatrix} 1 \\ 0 \\ 1 \end{pmatrix}$ is an eigenvector and $\begin{pmatrix} \frac{1}{\sqrt{2}} \\ 0 \\ \frac{1}{\sqrt{2}} \end{pmatrix}$ is the normalised eigenvector.

When $\lambda = 4$:

$$\begin{pmatrix} x + 2z \\ 4y \\ 2x + z \end{pmatrix} = \begin{pmatrix} 4x \\ 4y \\ 4z \end{pmatrix}$$

$$\Rightarrow \quad 2z = 3x$$
and
$$2x = 3z$$

which are inconsistent and imply $x = z = 0$.

So $\begin{pmatrix} 0 \\ 1 \\ 0 \end{pmatrix}$ is an eigenvector, which is already normalised.

Since
$$\begin{pmatrix} \frac{1}{\sqrt{2}} \\ 0 \\ -\frac{1}{\sqrt{2}} \end{pmatrix} \cdot \begin{pmatrix} \frac{1}{\sqrt{2}} \\ 0 \\ \frac{1}{\sqrt{2}} \end{pmatrix} = \tfrac{1}{2} - \tfrac{1}{2} = 0$$

$$\begin{pmatrix} \frac{1}{\sqrt{2}} \\ 0 \\ -\frac{1}{\sqrt{2}} \end{pmatrix} \cdot \begin{pmatrix} 0 \\ 1 \\ 0 \end{pmatrix} = 0$$

and
$$\begin{pmatrix} \frac{1}{\sqrt{2}} \\ 0 \\ \frac{1}{\sqrt{2}} \end{pmatrix} \cdot \begin{pmatrix} 0 \\ 1 \\ 0 \end{pmatrix} = 0$$

these eigenvectors are mutually orthogonal.

So:
$$\mathbf{P} = \begin{pmatrix} \frac{1}{\sqrt{2}} & \frac{1}{\sqrt{2}} & 0 \\ 0 & 0 & 1 \\ -\frac{1}{\sqrt{2}} & \frac{1}{\sqrt{2}} & 0 \end{pmatrix}$$

$$\mathbf{P}^{\mathsf{T}} = \begin{pmatrix} \frac{1}{\sqrt{2}} & 0 & -\frac{1}{\sqrt{2}} \\ \frac{1}{\sqrt{2}} & 0 & \frac{1}{\sqrt{2}} \\ 0 & 1 & 0 \end{pmatrix}$$

$$\mathbf{P}^{\mathsf{T}}\mathbf{AP} = \begin{pmatrix} \frac{1}{\sqrt{2}} & 0 & -\frac{1}{\sqrt{2}} \\ \frac{1}{\sqrt{2}} & 0 & \frac{1}{\sqrt{2}} \\ 0 & 1 & 0 \end{pmatrix} \begin{pmatrix} 1 & 0 & 2 \\ 0 & 4 & 0 \\ 2 & 0 & 1 \end{pmatrix} \begin{pmatrix} \frac{1}{\sqrt{2}} & \frac{1}{\sqrt{2}} & 0 \\ 0 & 0 & 1 \\ -\frac{1}{\sqrt{2}} & \frac{1}{\sqrt{2}} & 0 \end{pmatrix}$$

$$= \begin{pmatrix} \frac{1}{\sqrt{2}} & 0 & -\frac{1}{\sqrt{2}} \\ \frac{1}{\sqrt{2}} & 0 & \frac{1}{\sqrt{2}} \\ 0 & 1 & 0 \end{pmatrix} \begin{pmatrix} -\frac{1}{\sqrt{2}} & \frac{3}{\sqrt{2}} & 0 \\ 0 & 0 & 4 \\ \frac{1}{\sqrt{2}} & \frac{3}{\sqrt{2}} & 0 \end{pmatrix}$$

$$= \begin{pmatrix} -1 & 0 & 0 \\ 0 & 3 & 0 \\ 0 & 0 & 4 \end{pmatrix}$$

Exercise 3D

1 Find the eigenvalues and corresponding eigenvectors of the following matrices:

(a) $\begin{pmatrix} 2 & 4 \\ 5 & 3 \end{pmatrix}$

(b) $\begin{pmatrix} 1 & -2 \\ -5 & 4 \end{pmatrix}$

(c) $\begin{pmatrix} 2 & -1 \\ -8 & 4 \end{pmatrix}$

(d) $\begin{pmatrix} 3 & 1 \\ -1 & 1 \end{pmatrix}$

(e) $\begin{pmatrix} 3 & 2 \\ 2 & 3 \end{pmatrix}$

(f) $\begin{pmatrix} 1 & -\sqrt{2} \\ \sqrt{2} & 4 \end{pmatrix}$

(g) $\begin{pmatrix} 1 & 1 \\ 0 & 2 \end{pmatrix}$

2 Find the eigenvalues and corresponding eigenvectors of the following matrices:

(a) $\begin{pmatrix} -1 & 2 & 2 \\ 2 & 2 & 2 \\ -3 & -6 & -6 \end{pmatrix}$

(b) $\begin{pmatrix} 1 & 2 & 2 \\ 0 & 2 & 1 \\ -1 & 2 & 2 \end{pmatrix}$

(c) $\begin{pmatrix} 2 & 2 & 1 \\ 1 & 3 & 1 \\ 1 & 2 & 2 \end{pmatrix}$

(d) $\begin{pmatrix} 1 & 2 & 0 \\ 0 & 1 & 1 \\ 0 & 1 & 1 \end{pmatrix}$

(e) $\begin{pmatrix} 4 & 9 & 0 \\ 0 & -2 & 8 \\ 0 & 0 & 7 \end{pmatrix}$

(f) $\begin{pmatrix} 2 & 1 & 0 \\ -1 & 0 & 0 \\ 0 & -1 & -1 \end{pmatrix}$

(g) $\begin{pmatrix} 1 & 0 & -1 \\ 1 & 2 & 1 \\ 2 & 2 & 3 \end{pmatrix}$
(h) $\begin{pmatrix} 1 & -1 & 0 \\ 1 & 2 & 1 \\ -2 & 1 & -1 \end{pmatrix}$

(i) $\begin{pmatrix} 1 & 1 & 2 \\ 0 & 2 & 2 \\ -1 & 1 & 3 \end{pmatrix}$
(j) $\begin{pmatrix} 1 & 0 & 3 \\ 0 & 1 & -1 \\ 1 & -1 & 1 \end{pmatrix}$

3 Find an orthogonal matrix **P** such that $\mathbf{P}^{\mathrm{T}}\mathbf{A}\mathbf{P}$ diagonalises the symmetric matrix **A** where **A** =

(a) $\begin{pmatrix} 1 & 2 \\ 2 & 1 \end{pmatrix}$
(b) $\begin{pmatrix} 1 & -6 \\ -6 & -4 \end{pmatrix}$

(c) $\begin{pmatrix} 5 & -\sqrt{3} \\ -\sqrt{3} & 7 \end{pmatrix}$
(d) $\begin{pmatrix} 1 & -2 \\ -2 & 4 \end{pmatrix}$

(e) $\begin{pmatrix} 6 & -2 & 2 \\ -2 & 5 & 0 \\ 2 & 0 & 7 \end{pmatrix}$
(f) $\begin{pmatrix} 1 & 0 & 1 \\ 0 & 1 & 0 \\ 1 & 0 & 1 \end{pmatrix}$

(g) $\begin{pmatrix} 3 & -1 & -1 \\ -1 & 2 & 0 \\ -1 & 0 & 2 \end{pmatrix}$
(h) $\begin{pmatrix} 3 & -2 & -4 \\ -2 & -2 & -6 \\ -4 & -6 & -1 \end{pmatrix}$

(i) $\begin{pmatrix} 3 & -2 & -2 \\ -2 & 8 & -2 \\ -2 & -2 & 3 \end{pmatrix}$
(j) $\begin{pmatrix} 4 & -2 & 4 \\ -2 & 1 & -2 \\ 4 & -2 & 4 \end{pmatrix}$

(k) $\begin{pmatrix} 2 & -1 & 0 \\ -1 & 2 & -1 \\ 0 & -1 & 2 \end{pmatrix}$
(l) $\begin{pmatrix} 7 & -2 & -2 \\ -2 & 1 & 4 \\ -2 & 4 & 1 \end{pmatrix}$

(m) $\begin{pmatrix} 11 & 2 & 8 \\ 2 & 2 & -10 \\ 8 & -10 & 5 \end{pmatrix}$
(n) $\begin{pmatrix} 2 & -5 & 0 \\ -5 & -1 & 3 \\ 0 & 3 & -6 \end{pmatrix}$

SUMMARY OF KEY POINTS

1 A linear transformation T is such that

$$T(a_1\mathbf{v}_1 + a_2\mathbf{v}_2) = a_1 T(\mathbf{v}_1) + a_2 T(\mathbf{v}_2)$$

where a_1, a_2 are scalars and $\mathbf{v}_1$, $\mathbf{v}_2$ are vectors.

2 If T and S are linear transformations then TS is also a linear transformation and

$$TS(\mathbf{v}) = T[S(\mathbf{v})]$$

3 If a linear transformation T is one–one then an inverse linear transformation T^{-1} exists.

4 Matrix multiplication is *not* commutative. That is, in general,

$$\mathbf{A}\mathbf{B} \neq \mathbf{B}\mathbf{A}$$

5 The 2×2 identity matrix is $\mathbf{I} = \begin{pmatrix} 1 & 0 \\ 0 & 1 \end{pmatrix}$

The 3×3 identity matrix is $\mathbf{I} = \begin{pmatrix} 1 & 0 & 0 \\ 0 & 1 & 0 \\ 0 & 0 & 1 \end{pmatrix}$

6 If $\mathbf{A}\mathbf{B} = \mathbf{I}$, then $\mathbf{B} = \mathbf{A}^{-1}$ and $\mathbf{A} = \mathbf{B}^{-1}$.

7 $\begin{pmatrix} p & q \\ r & s \end{pmatrix}^{-1} = \dfrac{1}{ps - qr} \begin{pmatrix} s & -q \\ -r & p \end{pmatrix}$

8 If $\mathbf{A} = \begin{pmatrix} a & b & c \\ d & e & f \\ g & h & i \end{pmatrix}$

then the transpose of $\mathbf{A}$ is

$$\mathbf{A}^{\mathrm{T}} = \begin{pmatrix} a & d & g \\ b & e & h \\ c & f & i \end{pmatrix}$$

9 Det $\mathbf{A} = \begin{vmatrix} a & b & c \\ d & e & f \\ g & h & i \end{vmatrix} = a\begin{vmatrix} e & f \\ h & i \end{vmatrix} - b\begin{vmatrix} d & f \\ g & i \end{vmatrix} + c\begin{vmatrix} d & e \\ g & h \end{vmatrix}$

10 To find the inverse of a 3×3 matrix you
   (i)   form the matrix of minors
   (ii) form the matrix of cofactors
   (iii) transpose the matrix of cofactors
   (iv) divide the matrix of cofactors by the determinant.

11 A matrix is called singular if its determinant is zero.

12 Only non-singular square matrices have an inverse.

13 To find a 3×3 matrix which represents a linear transformation T, you find $T(\mathbf{i})$, $T(\mathbf{j})$ and $T(\mathbf{k})$ and make these the first, second and third columns respectively of the matrix.

14 If $\mathbf{A}$ and $\mathbf{B}$ are two matrices then

$$(\mathbf{A}\mathbf{B})^{-1} = \mathbf{B}^{-1}\mathbf{A}^{-1}$$

$$(\mathbf{A}\mathbf{B})^{\mathrm{T}} = \mathbf{B}^{\mathrm{T}}\mathbf{A}^{\mathrm{T}}$$

15 If $\mathbf{A}$ is a matrix and $\mathbf{v}$ is a non-zero vector such that $\mathbf{Av} = \lambda\mathbf{v}$ where λ is a scalar, then $\mathbf{v}$ is called an eigenvector of $\mathbf{A}$ and λ is the corresponding eigenvalue.

16 $|\mathbf{A} - \lambda\mathbf{I}| = 0$ is the characteristic equation of the matrix $\mathbf{A}$. The solutions of this equation give the eigenvalues of $\mathbf{A}$.

17 $\dfrac{1}{\sqrt{(a^2 + b^2 + c^2)}} \begin{pmatrix} a \\ b \\ c \end{pmatrix}$ is the normalised eigenvector of $\begin{pmatrix} a \\ b \\ c \end{pmatrix}$.

18 The eigenvectors $\mathbf{v}_1$ and $\mathbf{v}_2$ are orthogonal if

$$\mathbf{v}_1 . \mathbf{v}_2 = 0$$

19 The matrix $\begin{pmatrix} a & b & c \\ d & e & f \\ g & h & i \end{pmatrix}$ is orthogonal if $\begin{pmatrix} a \\ d \\ g \end{pmatrix}$, $\begin{pmatrix} b \\ e \\ h \end{pmatrix}$

and $\begin{pmatrix} c \\ f \\ i \end{pmatrix}$ are each normalised eigenvectors and if

$$\begin{pmatrix} a \\ d \\ g \end{pmatrix} . \begin{pmatrix} b \\ e \\ h \end{pmatrix} = 0, \quad \begin{pmatrix} a \\ d \\ g \end{pmatrix} . \begin{pmatrix} c \\ f \\ i \end{pmatrix} = 0 \text{ and } \begin{pmatrix} b \\ e \\ h \end{pmatrix} . \begin{pmatrix} c \\ f \\ i \end{pmatrix} = 0.$$

20 If $\mathbf{A}$ is orthogonal then $\mathbf{A}^T = \mathbf{A}^{-1}$.

21 A square matrix whose elements are all zero except those on the leading diagonal is called a diagonal matrix.

$\begin{pmatrix} a & 0 \\ 0 & b \end{pmatrix}$ and $\begin{pmatrix} a & 0 & 0 \\ 0 & b & 0 \\ 0 & 0 & c \end{pmatrix}$ are diagonal matrices.

22 A matrix $\mathbf{A}$ which has $\mathbf{A}^T = \mathbf{A}$ is symmetric.

23 If $\mathbf{A}$ is symmetric and $\mathbf{P}$ is an orthogonal matrix whose columns are the normalised, orthogonal eigenvectors of $\mathbf{A}$, then $\mathbf{P}^T\mathbf{AP}$ is diagonal.

Vectors

4

Chapter 5 of Book C4 introduced the subject of vectors and showed how they can be added, subtracted and multiplied by a scalar. It also described how to calculate the scalar product of two vectors and how to find an equation of a straight line in vector form. This chapter continues with vectors. It introduces another kind of product, the vector product, and shows you another form of the equation of a straight line. It concludes with cartesian and vector forms of the equation of a plane.

4.1 The vector product

The scalar product of the vectors **a** and **b** is

$$\mathbf{a.b} = |\mathbf{a}|\,|\mathbf{b}|\cos\theta$$

where θ is the angle between the vectors **a** and **b** (Book C4, page 64).

The **vector** (or **cross**) **product** of the vectors **a** and **b** is defined as

$$\mathbf{a} \times \mathbf{b} = |\mathbf{a}|\,|\mathbf{b}|\,\sin\theta\,\hat{\mathbf{n}}$$

Once again θ is the angle between **a** and **b**, and $\hat{\mathbf{n}}$ is a unit vector perpendicular to both **a** and **b**. The direction of $\hat{\mathbf{n}}$ is that in which a right-handed corkscrew would move when turned from **a** to **b**:

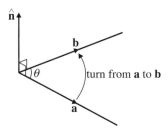

If the turn is in the opposite sense, that is from **b** to **a**, then the movement of the corkscrew is in the opposite sense to $\hat{\mathbf{n}}$. That is, it is in the direction of $-\hat{\mathbf{n}}$.

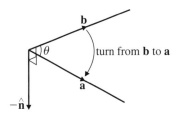

So:
$$\mathbf{b} \times \mathbf{a} = |\mathbf{b}|\,|\mathbf{a}| \sin\theta(-\hat{\mathbf{n}})$$
$$= -|\mathbf{a}|\,|\mathbf{b}| \sin\theta\,\hat{\mathbf{n}}$$
$$= -\mathbf{a} \times \mathbf{b}$$

■
$$\mathbf{b} \times \mathbf{a} = -\mathbf{a} \times \mathbf{b}$$

Be careful, therefore, because $\mathbf{a} \times \mathbf{b} \neq \mathbf{b} \times \mathbf{a}$. The vector product is *not* commutative. Notice that $\mathbf{a}.\mathbf{b}$ is called the scalar product of $\mathbf{a}$ and $\mathbf{b}$ because the result is a scalar and that $\mathbf{a} \times \mathbf{b}$ is called the vector product of $\mathbf{a}$ and $\mathbf{b}$ because the result is another vector.

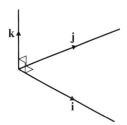

Now the vectors $\mathbf{i}$, $\mathbf{j}$ and $\mathbf{k}$ are such that each one is perpendicular to the other two. Also their relative positions are such that if a right-handed corkscrew were turned from $\mathbf{i}$ to $\mathbf{j}$ it would move in the direction $\mathbf{k}$. You should also be able to see that if a right-handed corkscrew were turned from $\mathbf{j}$ to $\mathbf{k}$ it would move in the direction $\mathbf{i}$ and if it were turned from $\mathbf{k}$ to $\mathbf{i}$, it would move in the direction $\mathbf{j}$.

So
$$\mathbf{i} \times \mathbf{j} = |\mathbf{i}|\,|\mathbf{j}| \sin 90°\,\mathbf{k}$$
$$= (1 \times 1 \times 1)\mathbf{k} = \mathbf{k}$$

■ That is: $\qquad\qquad \mathbf{i} \times \mathbf{j} = \mathbf{k}$

Similarly $\qquad\qquad \mathbf{j} \times \mathbf{k} = \mathbf{i}$

and $\qquad\qquad\quad \mathbf{k} \times \mathbf{i} = \mathbf{j}$

Also, since the angle between $\mathbf{i}$ and itself is 0 and $\sin 0 = 0$,

■ then: $\qquad\qquad\quad \mathbf{i} \times \mathbf{i} = \mathbf{0}$

Similarly $\qquad\qquad \mathbf{j} \times \mathbf{j} = \mathbf{0}$

and $\qquad\qquad\quad \mathbf{k} \times \mathbf{k} = \mathbf{0}$

The vector product of $a_1\mathbf{i} + a_2\mathbf{j} + a_3\mathbf{k}$ and $b_1\mathbf{i} + b_2\mathbf{j} + b_3\mathbf{k}$

$(a_1\mathbf{i} + a_2\mathbf{j} + a_3\mathbf{k}) \times (b_1\mathbf{i} + b_2\mathbf{j} + b_3\mathbf{k})$

$= a_1b_1(\mathbf{i} \times \mathbf{i}) + a_1b_2(\mathbf{i} \times \mathbf{j}) + a_1b_3(\mathbf{i} \times \mathbf{k}) + a_2b_1(\mathbf{j} \times \mathbf{i})$

$\quad + a_2b_2(\mathbf{j} \times \mathbf{j}) + a_2b_3(\mathbf{j} \times \mathbf{k}) + a_3b_1(\mathbf{k} \times \mathbf{i}) + a_3b_2(\mathbf{k} \times \mathbf{j}) + a_3b_3(\mathbf{k} \times \mathbf{k})$

$= a_1b_2\mathbf{k} + a_1b_3(-\mathbf{j}) + a_2b_1(-\mathbf{k}) + a_2b_3\mathbf{i} + a_3b_1\mathbf{j} + a_3b_2(-\mathbf{i})$

(because $\mathbf{i} \times \mathbf{k} = -\mathbf{k} \times \mathbf{i} = -\mathbf{j}, \mathbf{j} \times \mathbf{i} = -\mathbf{i} \times \mathbf{j} = -\mathbf{k}, \mathbf{k} \times \mathbf{j} = -\mathbf{j} \times \mathbf{k} = -\mathbf{i}$)

$= (a_2b_3 - a_3b_2)\mathbf{i} + (a_3b_1 - a_1b_3)\mathbf{j} + (a_1b_2 - a_2b_1)\mathbf{k}$

Now you should remember from chapter 3 (page 69) that the determinant

$$\begin{vmatrix} \mathbf{i} & \mathbf{j} & \mathbf{k} \\ a_1 & a_2 & a_3 \\ b_1 & b_2 & b_3 \end{vmatrix}$$

is $\qquad \mathbf{i}(a_2b_3 - a_3b_2) - \mathbf{j}(a_1b_3 - a_3b_1) + \mathbf{k}(a_1b_2 - a_2b_1)$

$\qquad = (a_2b_3 - a_3b_2)\mathbf{i} + (a_3b_1 - a_1b_3)\mathbf{j} + (a_1b_2 - a_2b_1)\mathbf{k}$

$\qquad = (a_1\mathbf{i} + a_2\mathbf{j} + a_3\mathbf{k}) \times (b_1\mathbf{i} + b_2\mathbf{j} + b_3\mathbf{k})$

So:

■ $\qquad (a_1\mathbf{i} + +a_2\mathbf{j} + a_3\mathbf{k}) \times (b_1\mathbf{i} + b_2\mathbf{j} + b_3\mathbf{k}) = \begin{vmatrix} \mathbf{i} & \mathbf{j} & \mathbf{k} \\ a_1 & a_2 & a_3 \\ b_1 & b_2 & b_3 \end{vmatrix}$

Vector product equal to zero

If $\mathbf{a} \times \mathbf{b} = \mathbf{0}$ then since $\mathbf{a} \times \mathbf{b} = |\mathbf{a}|\,|\mathbf{b}|\sin\theta\,\hat{\mathbf{n}}$

either $\qquad\qquad\qquad |\mathbf{a}| = 0 \Rightarrow \mathbf{a} = \mathbf{0}$

or $\qquad\qquad\qquad\quad |\mathbf{b}| = 0 \Rightarrow \mathbf{b} = \mathbf{0}$

or $\qquad\qquad\qquad\quad \sin\theta = 0 \Rightarrow \theta = 0$ or π

But if either $\theta = 0$ or $\theta = \pi$ then $\mathbf{a}$ and $\mathbf{b}$ are in the same direction (and either in the same sense or in opposite senses). In either case, if $\sin\theta = 0$, then $\mathbf{a}$ and $\mathbf{b}$ are parallel.

So:

■ if $\mathbf{a} \times \mathbf{b} = \mathbf{0}$ then either $\mathbf{a} = \mathbf{0}$, or $\mathbf{b} = \mathbf{0}$ or $\mathbf{a}$ and $\mathbf{b}$ are parallel.

Example 1

Simplify $(\mathbf{i} - 2\mathbf{j} + 5\mathbf{k}) \times (-2\mathbf{i} + \mathbf{j} - 3\mathbf{k})$ (a) directly (b) by evaluating an appropriate determinant.

(a) $(\mathbf{i} - 2\mathbf{j} + 5\mathbf{k}) \times (-2\mathbf{i} + \mathbf{j} - 3\mathbf{k})$

$= -2(\mathbf{i} \times \mathbf{i}) + (\mathbf{i} \times \mathbf{j}) - 3(\mathbf{i} \times \mathbf{k}) + 4(\mathbf{j} \times \mathbf{i}) - 2(\mathbf{j} \times \mathbf{j}) + 6(\mathbf{j} \times \mathbf{k}) - 10(\mathbf{k} \times \mathbf{i}) + 5(\mathbf{k} \times \mathbf{j}) - 15(\mathbf{k} \times \mathbf{k})$

$= \mathbf{0} + \mathbf{k} + 3\mathbf{j} - 4\mathbf{k} + \mathbf{0} + 6\mathbf{i} - 10\mathbf{j} - 5\mathbf{i} + \mathbf{0}$

$= \mathbf{i} - 7\mathbf{j} - 3\mathbf{k}$

(b) $\begin{vmatrix} \mathbf{i} & \mathbf{j} & \mathbf{k} \\ 1 & -2 & 5 \\ -2 & 1 & -3 \end{vmatrix}$

$= \mathbf{i}(6 - 5) - \mathbf{j}(-3 + 10) + \mathbf{k}(1 - 4)$

$= \mathbf{i} - 7\mathbf{j} - 3\mathbf{k}$

Example 2

Evaluate $(3\mathbf{i} - 2\mathbf{k}) \times (\mathbf{j} + 3\mathbf{k})$ (a) directly (b) by evaluating an appropriate determinant.

(a) $(3\mathbf{i} - 2\mathbf{k}) \times (\mathbf{j} + 3\mathbf{k})$

$= 3(\mathbf{i} \times \mathbf{j}) + 9(\mathbf{i} \times \mathbf{k}) - 2(\mathbf{k} \times \mathbf{j}) - 6(\mathbf{k} \times \mathbf{k})$

$= 3\mathbf{k} - 9\mathbf{j} + 2\mathbf{i} + \mathbf{0}$

$= 2\mathbf{i} - 9\mathbf{j} + 3\mathbf{k}$

(b) $\begin{vmatrix} \mathbf{i} & \mathbf{j} & \mathbf{k} \\ 3 & 0 & -2 \\ 0 & 1 & 3 \end{vmatrix}$

$= \mathbf{i}(0 + 2) - \mathbf{j}(9 - 0) + \mathbf{k}(3 - 0)$

$= 2\mathbf{i} - 9\mathbf{j} + 3\mathbf{k}$

As you can see from examples 1 and 2, you can usually save yourself a good deal of work if you evaluate a vector product by using a determinant.

Example 3

Find a unit vector which is perpendicular to both $\mathbf{a} = 2\mathbf{i} - \mathbf{j} + 3\mathbf{k}$ and $\mathbf{b} = -\mathbf{i} + 3\mathbf{j} - \mathbf{k}$.

The vector $\mathbf{a} \times \mathbf{b}$ is perpendicular to both $\mathbf{a}$ and $\mathbf{b}$. So a vector perpendicular to both $\mathbf{a}$ and $\mathbf{b}$ is given by

$$\begin{vmatrix} \mathbf{i} & \mathbf{j} & \mathbf{k} \\ 2 & -1 & 3 \\ -1 & 3 & -1 \end{vmatrix}$$

$$= \mathbf{i}(1 - 9) - \mathbf{j}(-2 + 3) + \mathbf{k}(6 - 1)$$

$$= -8\mathbf{i} - \mathbf{j} + 5\mathbf{k}$$

Since $\qquad |-8\mathbf{i} - \mathbf{j} + 5\mathbf{k}| = \sqrt{(64 + 1 + 25)} = \sqrt{90}$

a suitable unit vector is

$$\frac{1}{\sqrt{90}} (-8\mathbf{i} - \mathbf{j} + 5\mathbf{k})$$

Example 4

Find the sine of the acute angle between $\mathbf{a} = 2\mathbf{i} - \mathbf{j} + 2\mathbf{k}$ and $\mathbf{b} = -3\mathbf{i} + 4\mathbf{j} + \mathbf{k}$.

$$\mathbf{a} \times \mathbf{b} = |\mathbf{a}|\,|\mathbf{b}| \sin \theta\, \hat{\mathbf{n}}$$

where $\hat{\mathbf{n}}$ is a unit vector perpendicular to both $\mathbf{a}$ and $\mathbf{b}$.

Thus: $|\mathbf{a} \times \mathbf{b}| = \sqrt{(4 + 1 + 4)}\sqrt{(9 + 16 + 1)} \sin \theta$

Now $\qquad \mathbf{a} \times \mathbf{b} = \begin{vmatrix} \mathbf{i} & \mathbf{j} & \mathbf{k} \\ 2 & -1 & 2 \\ -3 & 4 & 1 \end{vmatrix}$

$$= \mathbf{i}(-1 - 8) - \mathbf{j}(2 + 6) + \mathbf{k}(8 - 3)$$

$$= -9\mathbf{i} - 8\mathbf{j} + 5\mathbf{k}$$

and $\qquad |\mathbf{a} \times \mathbf{b}| = \sqrt{(81 + 64 + 25)}$

So $\qquad \sin \theta = \dfrac{\sqrt{(81 + 64 + 25)}}{\sqrt{(4 + 1 + 4)}\sqrt{(9 + 16 + 1)}}$

$$= \frac{\sqrt{170}}{\sqrt{9}\,\sqrt{26}}$$

$$= \frac{\sqrt{170}}{3\sqrt{26}}$$

Thus $\qquad \sin \theta = \dfrac{\sqrt{85}}{3\sqrt{13}}$

Notice that you could alternatively have used the scalar product to find $\cos \theta$ and hence θ and $\sin \theta$ (see example 26 on page 67 of Book C4).

Exercise 4A

1 Simplify as much as possible:

(a) $3\mathbf{i} \times \mathbf{j}$

(b) $2\mathbf{i} \times (\mathbf{i} + \mathbf{j} + \mathbf{k})$

(c) $(\mathbf{i} + \mathbf{j}) \times (\mathbf{j} + \mathbf{k}) + (\mathbf{k} + \mathbf{i}) \times (\mathbf{j} - \mathbf{k})$

(d) $(\mathbf{i} + 2\mathbf{j} - \mathbf{k}) \times (2\mathbf{i} - \mathbf{j} + \mathbf{k})$

(e) $(2\mathbf{i} - 3\mathbf{j} + \mathbf{k}) \times (-\mathbf{i} + 2\mathbf{j} - 4\mathbf{k})$

(f) $(\mathbf{i} - \mathbf{j} + \mathbf{k}) \times (3\mathbf{i} - 3\mathbf{j} + 3\mathbf{k})$

(g) $(2\mathbf{i} + \mathbf{j} - 2\mathbf{k}) \times (-3\mathbf{i} + 4\mathbf{k})$

(h) $(2\mathbf{i} + \mathbf{k}) \times (\mathbf{i} - 2\mathbf{j} + 3\mathbf{k})$

(i) $(2\mathbf{i} + 3\mathbf{j} - \mathbf{k}) \times (2\mathbf{i} - \mathbf{j} + 3\mathbf{k})$

(j) $(-\mathbf{i} + 2\mathbf{j} - 3\mathbf{k}) \times (5\mathbf{i} - 4\mathbf{k})$

2 Find a unit vector which is perpendicular to the vector $(4\mathbf{i} + 4\mathbf{j} - 7\mathbf{k})$ and to the vector $(2\mathbf{i} + 2\mathbf{j} + \mathbf{k})$. **[E]**

3 Find a unit vector perpendicular to both $2\mathbf{i} - 6\mathbf{j} - 3\mathbf{k}$ and $4\mathbf{i} + 3\mathbf{j} - \mathbf{k}$.

4 Find a vector of magnitude 7 which is perpendicular to both $2\mathbf{i} + \mathbf{j} - 3\mathbf{k}$ and $\mathbf{i} - 2\mathbf{j} + \mathbf{k}$.

5 Find the magnitude of the vector $(\mathbf{i} + \mathbf{j} - \mathbf{k}) \times (\mathbf{i} - \mathbf{j} + \mathbf{k})$. **[E]**

6 Given that $\mathbf{a} = -\mathbf{i} + 2\mathbf{j} - 5\mathbf{k}$ and $\mathbf{b} = 5\mathbf{i} - 2\mathbf{j} + \mathbf{k}$ find

(a) $\mathbf{a} . \mathbf{b}$

(b) $\mathbf{a} \times \mathbf{b}$

(c) the unit vector in the direction of $\mathbf{a} \times \mathbf{b}$. **[E]**

7 Find the sine of the angle between $\mathbf{a}$ and $\mathbf{b}$ where

(a) $\mathbf{a} = 2\mathbf{i} - \mathbf{j}$, $\mathbf{b} = \mathbf{i} + \mathbf{j} - \mathbf{k}$

(b) $\mathbf{a} = \mathbf{i} + \mathbf{j} + 3\mathbf{k}$, $\mathbf{b} = -\mathbf{i} + 3\mathbf{k}$

(c) $\mathbf{a} = -2\mathbf{i} + \mathbf{j} + \mathbf{k}$, $\mathbf{b} = \mathbf{i} + 2\mathbf{j} + 2\mathbf{k}$

(d) $\mathbf{a} = \mathbf{i} - 2\mathbf{j} + 3\mathbf{k}$, $\mathbf{b} = 2\mathbf{i} - \mathbf{j} + 3\mathbf{k}$

(e) $\mathbf{a} = -\mathbf{i} - 2\mathbf{j} + \mathbf{k}$, $\mathbf{b} = 2\mathbf{i} + 3\mathbf{j} - \mathbf{k}$

8 Given that $\mathbf{a} = \mathbf{i} + 2\mathbf{j} - 2\mathbf{k}$ and $\mathbf{b} = p\mathbf{j} + q\mathbf{k}$ and that $\mathbf{a} \times \mathbf{b} = 2\mathbf{j} + \lambda\mathbf{k}$, find the values of the scalar constants p, q and λ.

9 Given that $\mathbf{u} = 2\mathbf{i} - \mathbf{j} + 2\mathbf{k}$, $\mathbf{v} = a\mathbf{i} + b\mathbf{k}$ and $\mathbf{u} \times \mathbf{v} = \mathbf{i} + c\mathbf{k}$, find the values of the scalar constants a, b and c. Find, in surd form, the cosine of the angle between $\mathbf{u}$ and $\mathbf{v}$.

10 Given that $\mathbf{r} = a\mathbf{i} + b\mathbf{j} + c\mathbf{k}$, $\mathbf{k} \times \mathbf{r} = \mathbf{p}$, $\mathbf{r} \times \mathbf{p} = \mathbf{k}$, where a, b, c are scalar constants, show that

$$a^2 + b^2 = 1 \quad \text{and} \quad c = 0$$

4.2 Applications of the vector product

Area of a triangle

You should know that if you are given a triangle ABC then the area of the triangle is given by $\frac{1}{2}ab\sin C$.

If, therefore, you have a triangle OAB, then the area of the triangle is given by

$$\tfrac{1}{2}(OA)(OB)\sin\theta$$

where

$$\angle AOB = \theta$$

Now if the position vectors of A and B relative to O are $\mathbf{a}$ and $\mathbf{b}$, respectively, then

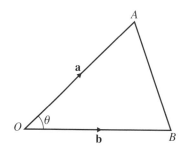

$$\text{area of } \triangle AOB = \tfrac{1}{2}(OA)(OB)\sin\theta$$

$$= \tfrac{1}{2}|\mathbf{a}|\,|\mathbf{b}|\sin\theta$$

$$= \tfrac{1}{2}|\mathbf{a}\times\mathbf{b}|$$

■ **Area of $\triangle AOB = \frac{1}{2}|\mathbf{a}\times\mathbf{b}|$**

If you have a triangle ABC where the position vectors of A, B, C relative to an origin O are $\mathbf{a}$, $\mathbf{b}$ and $\mathbf{c}$, respectively, then the area of the triangle can be calculated in a similar fashion.

The area of the triangle is given by $\frac{1}{2}(AB)(AC)\sin\theta$, where $\theta = \angle BAC$.

That is, $\frac{1}{2}|\overrightarrow{AB}|\,|\overrightarrow{AC}|\sin\theta$

But $\overrightarrow{AB} = \mathbf{b} - \mathbf{a}$ and $\overrightarrow{AC} = \mathbf{c} - \mathbf{a}$ so the area of the triangle is

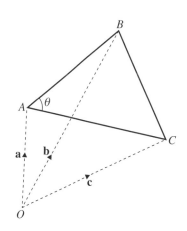

$$\tfrac{1}{2}|\mathbf{b} - \mathbf{a}|\,|\mathbf{c} - \mathbf{a}|\sin\theta$$

$$= \tfrac{1}{2}|(\mathbf{b} - \mathbf{a})\times(\mathbf{c} - \mathbf{a})|$$

$$= \tfrac{1}{2}|(\mathbf{b}\times\mathbf{c}) - (\mathbf{b}\times\mathbf{a}) - (\mathbf{a}\times\mathbf{c}) + (\mathbf{a}\times\mathbf{a})|$$

$$= \tfrac{1}{2}|\mathbf{b}\times\mathbf{c} + \mathbf{c}\times\mathbf{a} + \mathbf{a}\times\mathbf{b}|$$

$$= \tfrac{1}{2}|\mathbf{a}\times\mathbf{b} + \mathbf{b}\times\mathbf{c} + \mathbf{c}\times\mathbf{a}|$$

since $\mathbf{c}\times\mathbf{a} = -\mathbf{a}\times\mathbf{c}$, $\mathbf{a}\times\mathbf{b} = -\mathbf{b}\times\mathbf{a}$ and $\mathbf{a}\times\mathbf{a} = \mathbf{0}$.

■ **Area of $\triangle ABC = \frac{1}{2}|\mathbf{a}\times\mathbf{b} + \mathbf{b}\times\mathbf{c} + \mathbf{c}\times\mathbf{a}|$**

Example 5
Find the area of the triangle OAB where O is the origin, A has position vector $2\mathbf{i} - \mathbf{j} + \mathbf{k}$ and B has position vector $3\mathbf{i} + 4\mathbf{j} - 2\mathbf{k}$.

The area of the triangle is given by

$$\tfrac{1}{2}|(2\mathbf{i} - \mathbf{j} + \mathbf{k}) \times (3\mathbf{i} + 4\mathbf{j} - 2\mathbf{k})|$$

Now
$$(2\mathbf{i} - \mathbf{j} + \mathbf{k}) \times (3\mathbf{i} + 4\mathbf{j} - 2\mathbf{k})$$

$$= \begin{vmatrix} \mathbf{i} & \mathbf{j} & \mathbf{k} \\ 2 & -1 & 1 \\ 3 & 4 & -2 \end{vmatrix}$$

$$= \mathbf{i}(2 - 4) - \mathbf{j}(-4 - 3) + \mathbf{k}(8 + 3)$$

$$= -2\mathbf{i} + 7\mathbf{j} + 11\mathbf{k}$$

So the area of the triangle is

$$\tfrac{1}{2}\sqrt{(4 + 49 + 121)} = \tfrac{1}{2}\sqrt{174}$$

Example 6
Find the area of the triangle ABC where the position vectors of A, B, C relative to the origin O are $2\mathbf{i} + 5\mathbf{j} - \mathbf{k}$, $3\mathbf{i} - 4\mathbf{j} + 2\mathbf{k}$ and $-\mathbf{i} + 2\mathbf{j} - \mathbf{k}$ respectively.

$$\overrightarrow{AB} = \overrightarrow{OB} - \overrightarrow{OA} = \mathbf{i} - 9\mathbf{j} + 3\mathbf{k}$$

$$\overrightarrow{AC} = \overrightarrow{OC} - \overrightarrow{OA} = -3\mathbf{i} - 3\mathbf{j}$$

Now
$$\overrightarrow{AB} \times \overrightarrow{AC} = \begin{vmatrix} \mathbf{i} & \mathbf{j} & \mathbf{k} \\ 1 & -9 & 3 \\ -3 & -3 & 0 \end{vmatrix}$$

$$= \mathbf{i}(0 + 9) - \mathbf{j}(0 + 9) + \mathbf{k}(-3 - 27)$$

$$= 9\mathbf{i} - 9\mathbf{j} - 30\mathbf{k}$$

So:
$$\text{Area of } \triangle ABC = \tfrac{1}{2}|\overrightarrow{AB} \times \overrightarrow{AC}|$$

$$= \tfrac{1}{2}\sqrt{(81 + 81 + 900)}$$

$$= \tfrac{1}{2} \times 3\sqrt{(9 + 9 + 100)}$$

$$= \tfrac{3}{2}\sqrt{(118)}$$

Volume of a parallelepiped

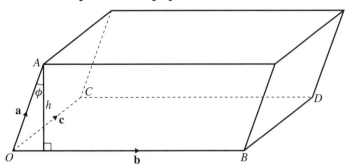

The volume of a parallelepiped is given by (area of base) $\times h$ where h is the perpendicular distance between the base and the top face. In the parallelepiped above, O is the origin and A, B, C have position vectors $\mathbf{a}$, $\mathbf{b}$ and $\mathbf{c}$ respectively. The base of the parallelepiped is a parallelogram.

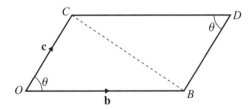

Now the area of $\triangle OBC$ is $\frac{1}{2}(OB)(OC)\sin\theta$ and the area of $\triangle DCB$ is $\frac{1}{2}(DC)(DB)\sin\theta$. (You should know that the opposite angles of a parallelogram are equal in size.) But because $OBDC$ is a parallelogram, $OC = BD$ and $OB = CD$.

So:

$$\text{area of parallelogram } OBDC = \tfrac{1}{2}(OB)(OC)\sin\theta + \tfrac{1}{2}(DC)(DB)\sin\theta$$

$$= \tfrac{1}{2}(OB)(OC)\sin\theta + \tfrac{1}{2}(OB)(OC)\sin\theta$$

$$= (OB)(OC)\sin\theta$$

$$= |\mathbf{b} \times \mathbf{c}|$$

The volume of the parallelepiped is therefore $|\mathbf{b} \times \mathbf{c}|h$.

Now
$$\frac{h}{OA} = \cos\phi$$

where ϕ is the angle between the vertical and OA.

So:
$$h = OA\cos\phi$$

and the volume is
$$|\mathbf{b} \times \mathbf{c}|OA\cos\phi$$

$$= |\mathbf{b} \times \mathbf{c}|\,|\mathbf{a}|\cos\phi$$

$$= |\mathbf{a}|\,|\mathbf{b} \times \mathbf{c}|\cos\phi$$

But $\mathbf{b} \times \mathbf{c}$ is vertically up, in the direction of h, since $\mathbf{b} \times \mathbf{c}$ is perpendicular to both $\mathbf{b}$ and $\mathbf{c}$.

So ϕ is the angle between $\mathbf{a}$ and $\mathbf{b} \times \mathbf{c}$, and $|\mathbf{a}| \, |\mathbf{b} \times \mathbf{c}| \cos \phi$ is the scalar product of $\mathbf{a}$ and $\mathbf{b} \times \mathbf{c}$.

■ **Thus the volume of the parallelepiped is given by $\mathbf{a}.(\mathbf{b} \times \mathbf{c})$, which is usually written without the brackets, because there can be no confusion, as simply $\mathbf{a}.\mathbf{b} \times \mathbf{c}$. This quantity is usually known as a triple scalar product.**

Evaluating the triple scalar product

You know that $\mathbf{b} \times \mathbf{c}$ can be evaluated as

$$\begin{vmatrix} \mathbf{i} & \mathbf{j} & \mathbf{k} \\ b_1 & b_2 & b_3 \\ c_1 & c_2 & c_3 \end{vmatrix}$$

where $\qquad\qquad\qquad \mathbf{b} = b_1\mathbf{i} + b_2\mathbf{j} + b_3\mathbf{k}$

and $\qquad\qquad\qquad \mathbf{c} = c_1\mathbf{i} + c_2\mathbf{j} + c_3\mathbf{k}$

So $\quad \mathbf{b} \times \mathbf{c} = (b_2c_3 - b_3c_2)\mathbf{i} + (b_3c_1 - b_1c_3)\mathbf{j} + (b_1c_2 - b_2c_1)\mathbf{k}$ $\quad$ (A)

If $\mathbf{a} = a_1\mathbf{i} + a_2\mathbf{j} + a_3\mathbf{k}$, then:

$$\mathbf{a}.\mathbf{b} \times \mathbf{c} = a_1(b_2c_3 - b_3c_2) + a_2(b_3c_1 - b_1c_3) + a_3(b_1c_2 - b_2c_1) \quad \text{(B)}$$

(See Book P3, page 163.)

However, if you compare (A) and (B) you will see that they are the same, except that in (B) $\mathbf{i}$ is replaced by a_1, $\mathbf{j}$ is replaced by a_2 and $\mathbf{k}$ is replaced by a_3. This leads to the conclusion that

■ $\qquad\qquad \mathbf{a}.\mathbf{b} \times \mathbf{c} = \begin{vmatrix} a_1 & a_2 & a_3 \\ b_1 & b_2 & b_3 \\ c_1 & c_2 & c_3 \end{vmatrix}$

Volume of a tetrahedron

The volume of a tetrahedron is given by the formula $\frac{1}{3}$ (area of base) $\times h$, where h is the perpendicular height.

In the tetrahedron $OABC$, you have O as the origin, $\mathbf{a}$ as the position vector of A, $\mathbf{b}$ as the position vector of B and $\mathbf{c}$ as the position vector of C. The perpendicular height makes an angle ϕ with OA.

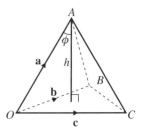

You know that the area of the triangular base is $\frac{1}{2}|\mathbf{b} \times \mathbf{c}|$.

Also: $\qquad\qquad\qquad h = OA \cos \phi = |\mathbf{a}| \cos \phi$

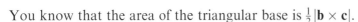

So the volume of the tetrahedron is given by

$$\tfrac{1}{3} \times \tfrac{1}{2} |\mathbf{b} \times \mathbf{c}| \, |\mathbf{a}| \cos \phi$$

But because $\mathbf{b} \times \mathbf{c}$ is in the direction of h, the angle ϕ is the angle between $\mathbf{b} \times \mathbf{c}$ and $\mathbf{a}$.

■ **Therefore the volume of the tetrahedron is given by**

$$\tfrac{1}{6} \mathbf{a}.\mathbf{b} \times \mathbf{c}$$

Example 7

A tetrahedron has vertices at $A(0, 1, 0)$, $B(1, 1, 2)$, $C(-2, 1, 3)$ and $D(2, 0, 1)$ relative to the origin O. Find the volume of the tetrahedron.

$$\overrightarrow{AB} = \overrightarrow{OB} - \overrightarrow{OA} = \mathbf{i} + 2\mathbf{k}$$

$$\overrightarrow{AC} = \overrightarrow{OC} - \overrightarrow{OA} = -2\mathbf{i} + 3\mathbf{k}$$

$$\overrightarrow{AD} = \overrightarrow{OD} - \overrightarrow{OA} = 2\mathbf{i} - \mathbf{j} + \mathbf{k}$$

The volume of the tetrahedron is given by

$$\tfrac{1}{6} \overrightarrow{AB}.\overrightarrow{AC} \times \overrightarrow{AD} = \tfrac{1}{6} \begin{vmatrix} 1 & 0 & 2 \\ -2 & 0 & 3 \\ 2 & -1 & 1 \end{vmatrix}$$

$$= \tfrac{1}{6}[1(3) + 0 + 2(2)]$$

$$= \tfrac{7}{6}$$

You should notice that sometimes when you are using the triple scalar product to evaluate a volume the answer is negative. This is because when you evaluate the vector product to find the area of the base of the figure you may unwittingly evaluate the product in the reverse order (that is, you evaluate $\mathbf{c} \times \mathbf{b}$ instead of $\mathbf{b} \times \mathbf{c}$). The result is therefore a vector pointing in the opposite direction to the perpendicular height of the figure. You should remember from page 110 that $\mathbf{c} \times \mathbf{b} = -(\mathbf{b} \times \mathbf{c})$. This is what causes the answer to be negative. Under these circumstances, the answer you require is the *modulus* of the triple scalar product.

Exercise 4B

1 Find the area of the triangle with vertices $A(0, 0, 0)$, $B(1, 2, 1)$ and $C(-1, 3, 3)$.

2 Find the area of the triangle with vertices $A(-5, 1, 4)$, $B(0, 0, 0)$ and $C(-2, 3, -1)$.

3 Find the area of the triangle with vertices $A(1, -2, 3)$, $B(-1, -1, 4)$ and $C(-2, 1, 5)$.

4 Find the area of the triangle with vertices $A(2, -1, -1)$, $B(-2, 1, -3)$ and $C(1, -1, 0)$.

5 Find the area of the triangle with vertices $A(-1, 3, 1)$, $B(2, 2, -3)$ and $C(-1, 3, -4)$.

6 Find the area of the parallelogram $ABCD$ where A is the point with coordinates $(1, 2, -3)$, B is the point with coordinates $(-1, 3, -4)$ and D is the point with coordinates $(1, 5, -2)$.

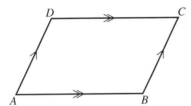

7 Find the area of the parallelogram $ABCD$ in which the vertices A, B and D have coordinates $(-1, 2, 1)$, $(3, 1, 2)$ and $(5, 1, -6)$ respectively.

8 Find the area of the triangle with vertices $A(3, -1, 2)$, $B(1, -1, 3)$ and $C(4, -3, 1)$.

9

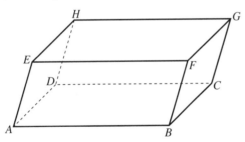

Find the volume of the parallelepiped $ABCDEFGH$ where the vertices A, B, D and E have coordinates $(0, 0, 0)$, $(5, -2, 3)$, $(2, -3, 4)$ and $(3, -1, -2)$ respectively.

10 The points A, B, C, D have position vectors

$$\mathbf{a} = (2\mathbf{i} + \mathbf{j}) \qquad \mathbf{b} = (3\mathbf{i} - \mathbf{j} + \mathbf{k})$$
$$\mathbf{c} = (-2\mathbf{j} - \mathbf{k}) \qquad \mathbf{d} = (2\mathbf{i} - \mathbf{j} + 3\mathbf{k})$$

respectively.
(a) Find $\overrightarrow{AB} \times \overrightarrow{BC}$ and $\overrightarrow{BD} \times \overrightarrow{DC}$.
(b) Hence find
    (i) the area of $\triangle ABC$
    (ii) the volume of the tetrahedron $ABCD$.   [E]

11 Relative to an origin O, the points P and Q have position vectors **p** and **q** respectively, where

$$\mathbf{p} = a(\mathbf{i} + \mathbf{j} + 2\mathbf{k}) \quad \text{and} \quad \mathbf{q} = a(2\mathbf{i} + \mathbf{j} + 3\mathbf{k}) \text{ and } a > 0.$$

Find the area of triangle OPQ. [E]

12 Referred to an origin O, the points P and Q have position vectors $3\mathbf{i} - 3\mathbf{k}$ and $\mathbf{i} + 2\mathbf{j} - 7\mathbf{k}$ respectively. Find
(a) $\overrightarrow{OP}.\overrightarrow{OQ}$
(b) $\overrightarrow{OP} \times \overrightarrow{OQ}$
(c) the size, in degrees to $0.1°$, of $\angle POQ$
(d) the area of $\triangle OPQ$.

13 Referred to O as origin, $\overrightarrow{OA} = 3\mathbf{i} + \mathbf{j} - 2\mathbf{k}$, $\overrightarrow{OB} = \mathbf{i} - 2\mathbf{j} + 3\mathbf{k}$ and $\overrightarrow{OC} = -\mathbf{i} + 4\mathbf{j} + 2\mathbf{k}$.
(a) Show that $\overrightarrow{AB}$ is perpendicular to $\overrightarrow{OC}$.
(b) Find the area of $\triangle OAB$.
(c) Calculate the area of $\triangle ABC$.

14 The points $A(1, -1, -1)$, $B(-1, 1, -1)$, $C(-1, -1, 1)$ and $D(1, 1, 1)$ are given referred to a fixed origin O.
(a) Show that $ABCD$ is a regular tetrahedron.
(b) Find the volume of $ABCD$.

15 Find the volume of the tetrahedron with vertices at the points $(1, 3, -1)$, $(2, 2, 3)$, $(4, 2, -2)$ and $(3, 7, 4)$.

16 A tetrahedron has its vertices at the points $O(0, 0, 0)$, $A(-1, 1, 2)$, $B(1, 2, -1)$ and $C(0, 1, 3)$.
(a) Determine the area of the face ABC.
(b) Find a unit vector normal to the face ABC.
(c) Find the volume of the tetrahedron.

17 Find the volume of the tetrahedron with vertices $(0, 1, 0)$, $(0, 0, -4)$, $(2, -1, 3)$, $(2, -1, 2)$.

18 The tetrahedron $ABCD$ has vertices $A(1, -1, 0)$, $B(0, 2, -1)$, $C(0, 2, 1)$, $D(-1, 3, 0)$.
(a) Find the area of face BCD.
(b) Find the volume of the tetrahedron.

19 A tetrahedron $OABC$ has its vertices at the points $O(0, 0, 0)$, $A(1, 2, -1)$, $B(-1, 1, 2)$ and $C(2, -1, 1)$.

(a) Write down expressions for $\overrightarrow{AB}$ and $\overrightarrow{AC}$ in terms of $\mathbf{i}$, $\mathbf{j}$ and $\mathbf{k}$ and find $\overrightarrow{AB} \times \overrightarrow{AC}$.

(b) Deduce the area of $\triangle ABC$.

(c) Find the volume of the tetrahedron $OABC$.   [E]

20 The edges OP, OQ, OR of a tetrahedron $OPQR$ are the vectors $\mathbf{a}$, $\mathbf{b}$ and $\mathbf{c}$ respectively, where

$$\mathbf{a} = 2\mathbf{i} + 4\mathbf{j}$$
$$\mathbf{b} = 2\mathbf{i} - \mathbf{j} + 3\mathbf{k}$$
$$\mathbf{c} = 4\mathbf{i} - 2\mathbf{j} + 5\mathbf{k}$$

(a) Evaluate $\mathbf{b} \times \mathbf{c}$ and deduce that OP is perpendicular to the plane OQR.

(b) Write down the length of OP and the area of $\triangle OQR$ and hence the volume of the tetrahedron.

(c) Verify your result by evaluating $\mathbf{a}.(\mathbf{b} \times \mathbf{c})$.   [E]

4.3 The equation of a straight line

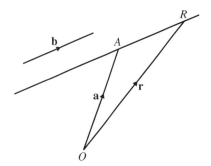

If a line is parallel to a vector $\mathbf{b}$, and if a point A on the line has position vector $\mathbf{a}$ and any other point R on the line has position vector $\mathbf{r}$, then an equation of the line is

$$\mathbf{r} = \mathbf{a} + \lambda\mathbf{b}$$

where λ is a scalar parameter (Book C4, chapter 5).

Now the vector $\overrightarrow{AR} = \mathbf{r} - \mathbf{a}$ and $\overrightarrow{AR}$ is parallel to $\mathbf{b}$. But you learned earlier (page 111) that if two vectors are parallel, then their vector product is zero.

■ **So another form of the equation of the straight line is**
$$(\mathbf{r} - \mathbf{a}) \times \mathbf{b} = 0$$

Example 8

Find an equation of the straight line which passes through point $A(2, -1, 1)$ and is parallel to $\mathbf{i} + 4\mathbf{j} - 3\mathbf{k}$.

The position vector of A is $2\mathbf{i} - \mathbf{j} + \mathbf{k}$. If R is any point on the line with position vector $\mathbf{r}$ then

$$\overrightarrow{AR} = \mathbf{r} - (2\mathbf{i} - \mathbf{j} + \mathbf{k})$$

and this is parallel to $\mathbf{i} + 4\mathbf{j} - 3\mathbf{k}$.

So: $$[\mathbf{r} - (2\mathbf{i} - \mathbf{j} + \mathbf{k})] \times (\mathbf{i} + 4\mathbf{j} - 3\mathbf{k}) = \mathbf{0}$$

or $$\mathbf{r} \times (\mathbf{i} + 4\mathbf{j} - 3\mathbf{k}) = (2\mathbf{i} - \mathbf{j} + \mathbf{k}) \times (\mathbf{i} + 4\mathbf{j} - 3\mathbf{k})$$

Thus $$\mathbf{r} \times (\mathbf{i} + 4\mathbf{j} - 3\mathbf{k}) = -\mathbf{i} + 7\mathbf{j} + 9\mathbf{k}$$

is an equation of the line and is equivalent to the equation

$$\mathbf{r} = (2\mathbf{i} - \mathbf{j} + \mathbf{k}) + \lambda(\mathbf{i} + 4\mathbf{j} - 3\mathbf{k})$$

which you already know.

4.4 The scalar product form of the equation of a plane

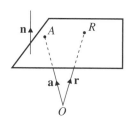

The point A, with position vector $\mathbf{a}$ lies in a given plane. The vector $\mathbf{n}$ is perpendicular to the plane. The point R with position vector $\mathbf{r}$ is any other point in the plane.

Then: $$\overrightarrow{AR} = \mathbf{r} - \mathbf{a}$$

Since both A and R lie in the given plane, then $\overrightarrow{AR}$ must lie in the plane. But if $\mathbf{n}$ is perpendicular to the plane then the directions of $\mathbf{n}$ and $\overrightarrow{AR} = \mathbf{r} - \mathbf{a}$ must be perpendicular. However, if two vectors are perpendicular then their scalar product is zero (see Book C4, page 64).

So: $$(\mathbf{r} - \mathbf{a}).\mathbf{n} = 0$$

or $$\mathbf{r}.\mathbf{n} - \mathbf{a}.\mathbf{n} = 0$$

$$\Rightarrow \quad \mathbf{r}.\mathbf{n} = \mathbf{a}.\mathbf{n}$$

If $\mathbf{a}.\mathbf{n} = p$ then an equation of the plane is $\mathbf{r}.\mathbf{n} = p$.

■ **Given that a plane contains a point with position vector a, that r is the position vector of any other point in the plane and that n is perpendicular to the plane, the scalar product form of the equation of the plane is r . n = p, where p = a . n.**

Example 9

The point A with position vector $\mathbf{i} - 2\mathbf{j} + \mathbf{k}$ lies in a plane. The vector $-\mathbf{i} + \mathbf{j} - \mathbf{k}$ is perpendicular to the plane. Find an equation of the plane
  (a) in scalar product form
  (b) in cartesian form.

(a) If R is any point in the plane and has position vector $\mathbf{r}$ then

$$[\mathbf{r} - (\mathbf{i} - 2\mathbf{j} + \mathbf{k})].(-\mathbf{i} + \mathbf{j} - \mathbf{k}) = 0$$
$$\mathbf{r}.(-\mathbf{i} + \mathbf{j} - \mathbf{k}) = (\mathbf{i} - 2\mathbf{j} + \mathbf{k}).(-\mathbf{i} + \mathbf{j} - \mathbf{k})$$
$$\mathbf{r}.(-\mathbf{i} + \mathbf{j} - \mathbf{k}) = -1 - 2 - 1$$

So $\mathbf{r}.(-\mathbf{i} + \mathbf{j} - \mathbf{k}) = -4$ is an equation of the plane in scalar product form.

(b) If $\mathbf{r} = x\mathbf{i} + y\mathbf{j} + z\mathbf{k}$

then: $$(x\mathbf{i} + y\mathbf{j} + z\mathbf{k}).(-\mathbf{i} + \mathbf{j} - \mathbf{k}) = -4$$

$$\Rightarrow \quad -x + y - z = -4$$

or: $$x - y + z = 4$$

which is a cartesian equation of the plane.

Example 10

Show that the line with equation $\mathbf{r} = \mathbf{i} + 2\mathbf{j} + \mathbf{k} + \lambda(2\mathbf{i} + \mathbf{j})$ where λ is a scalar parameter lies in the plane with equation $\mathbf{r}.(\mathbf{i} - 2\mathbf{j} + 2\mathbf{k}) = -1$.

The line passes through the point with position vector $\mathbf{i} + 2\mathbf{j} + \mathbf{k}$.

Now: $$(\mathbf{i} + 2\mathbf{j} + \mathbf{k}).(\mathbf{i} - 2\mathbf{j} + 2\mathbf{k}) = 1 - 4 + 2 = -1$$

The point with position vector $\mathbf{i} + 2\mathbf{j} + \mathbf{k}$ therefore satisfies the equation of the plane $\mathbf{r}.(\mathbf{i} - 2\mathbf{j} + 2\mathbf{k}) = -1$. Thus the point with position vector $\mathbf{i} + 2\mathbf{j} + \mathbf{k}$ lies on the line and also lies in the plane.

If you take $\lambda = 1$, then $\mathbf{r} = 3\mathbf{i} + 3\mathbf{j} + \mathbf{k}$ lies on the line. Using the equation of the plane you have

$$(3\mathbf{i} + 3\mathbf{j} + \mathbf{k}).(\mathbf{i} - 2\mathbf{j} + 2\mathbf{k}) = 3 - 6 + 2 = -1$$

So this point is also on the line and the plane, and hence the line lies in the plane.

4.5 The vector equation of a plane

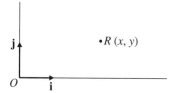

You know that if you have two perpendicular vectors **i** and **j** you can obtain the position vector of any other point in their plane by taking a number of **i**'s and combining these with a number of **j**'s. So the position vector of R is $x\mathbf{i} + y\mathbf{j}$.

Now the two vectors you start with do not have to be unit vectors, nor do they have to be at right angles to each other. So long as neither of the vectors is the zero vector and so long as the two vectors are not parallel to each other then you can still obtain the position vector of any point in the plane. (If the two vectors are parallel you will only be able to get points along a straight line in the direction of these vectors.)

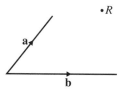

So to get to any point R which lies in the same plane as **a** and **b** you just need a number of **a**'s together with a number of **b**'s. That is, if **r** is the position vector of R then

$$\mathbf{r} = \lambda\mathbf{a} + \mu\mathbf{b}$$

where λ and μ are scalar parameters.

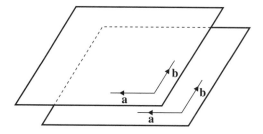

This is all very well, but if you have a number of parallel planes then the direction vectors **a** and **b** will be common to each of them. So if you have $\lambda\mathbf{a} + \mu\mathbf{b}$ how do you know in which one of the parallel planes $\lambda\mathbf{a} + \mu\mathbf{b}$ lies? Well, you don't of course! However, if you can find the position vector of just one point which lies in your plane then that plane can be defined uniquely.

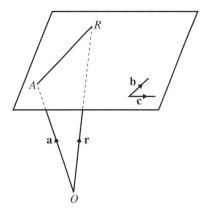

If A lies in the given plane and has position vector $\mathbf{a}$ and R is any point in the plane and has position vector $\mathbf{r}$, and $\mathbf{b}$ and $\mathbf{c}$ are two non-parallel vectors in the plane, neither of which is zero, then

$$\mathbf{r} = \mathbf{a} + \overrightarrow{AR}$$

But $\overrightarrow{AR}$, since it lies in the plane, can be written:

$$\overrightarrow{AR} = \lambda\mathbf{b} + \mu\mathbf{c}$$

where λ, μ are scalar parameters.

So:
$$\mathbf{r} = \mathbf{a} + \lambda\mathbf{b} + \mu\mathbf{c}$$

■ **The vector equation of a plane where a is the position vector of a point in the plane and b and c are non-parallel vectors in the plane, neither of which is zero, is given by**

$$\mathbf{r} = \mathbf{a} + \lambda\mathbf{b} + \mu\mathbf{c},$$

where λ and μ are scalars.

Example 11
Three points in a plane have coordinates $A(1, -1, 0)$, $B(0, 1, 1)$ and $C(2, 1, -2)$ referred to an origin O. Find a vector equation of the plane.

$$\overrightarrow{AB} = \overrightarrow{OB} - \overrightarrow{OA} = -\mathbf{i} + 2\mathbf{j} + \mathbf{k}$$
$$\overrightarrow{AC} = \overrightarrow{OC} - \overrightarrow{OA} = \mathbf{i} + 2\mathbf{j} - 2\mathbf{k}$$

Since $\mathbf{i} - \mathbf{j}$ is the position vector of A, an equation of the plane is:

$$\mathbf{r} = \mathbf{i} - \mathbf{j} + \lambda(-\mathbf{i} + 2\mathbf{j} + \mathbf{k}) + \mu(\mathbf{i} + 2\mathbf{j} - 2\mathbf{k})$$

Example 12
Find a cartesian equation of the plane containing the points $A(1, 1, 1)$, $B(2, 1, 0)$ and $C(2, 2, -1)$.

$$\overrightarrow{AB} = \mathbf{i} - \mathbf{k}, \quad \overrightarrow{AC} = \mathbf{i} + \mathbf{j} - 2\mathbf{k}$$

$$\overrightarrow{AB} \times \overrightarrow{AC} = \begin{vmatrix} \mathbf{i} & \mathbf{j} & \mathbf{k} \\ 1 & 0 & -1 \\ 1 & 1 & -2 \end{vmatrix} = \mathbf{i} + \mathbf{j} + \mathbf{k}$$

and is normal to the plane.

So an equation of the plane is

$$[\mathbf{r} - (2\mathbf{i} + \mathbf{j})].(\mathbf{i} + \mathbf{j} + \mathbf{k}) = 0$$

since the point B which is in the plane has position vector $2\mathbf{i} + \mathbf{j}$.

That is: $\qquad \mathbf{r}.(\mathbf{i} + \mathbf{j} + \mathbf{k}) = (2\mathbf{i} + \mathbf{j}).(\mathbf{i} + \mathbf{j} + \mathbf{k}) = 3$

or $\qquad\qquad\qquad\qquad\qquad x + y + z = 3$

Exercise 4C

1 Find, in the form $(\mathbf{r} - \mathbf{a}) \times \mathbf{b} = \mathbf{0}$, an equation of the straight line passing through the point with position vector $\mathbf{a}$ and which is parallel to the vector $\mathbf{b}$ where:
 (a) $\mathbf{a} = 2\mathbf{i} - \mathbf{j} + \mathbf{k}, \quad \mathbf{b} = -\mathbf{i} + \mathbf{j} - 2\mathbf{k}$
 (b) $\mathbf{a} = \mathbf{i} - 2\mathbf{k}, \quad \mathbf{b} = 2\mathbf{i} + 3\mathbf{j}$
 (c) $\mathbf{a} = \mathbf{i} - 5\mathbf{j} + 2\mathbf{k}, \quad \mathbf{b} = 2\mathbf{i} + \mathbf{j} - 3\mathbf{k}$
 (d) $\mathbf{a} = \mathbf{i} - \mathbf{j} + 4\mathbf{k}, \quad \mathbf{b} = 3\mathbf{i} - 2\mathbf{j} + \mathbf{k}$

2 Find, in the form $(\mathbf{r} - \mathbf{a}) \times \mathbf{b} = \mathbf{0}$, an equation of the line passing through the points with coordinates
 (a) $(3, 2, 5)$, $(9, 5, 17)$
 (b) $(1, -2, -1)$, $(2, -4, -4)$
 (c) $(-4, -3, 11)$, $(0, -3, 1)$
 (d) $(3, -2, 3)$, $(-1, 0, 1)$

3 Find an equation, in the form $(\mathbf{r} - \mathbf{a}) \times \mathbf{b} = \mathbf{0}$, of the line given by the equation, where λ is a scalar:
 (a) $\mathbf{r} = \mathbf{i} - \mathbf{j} + 2\mathbf{k} + \lambda(2\mathbf{i} - \mathbf{j} + \mathbf{k})$
 (b) $\mathbf{r} = -\mathbf{i} + 3\mathbf{j} + \mathbf{k} + \lambda(-\mathbf{i} - 3\mathbf{j} - 4\mathbf{k})$
 (c) $\mathbf{r} = \mathbf{i} - 4\mathbf{k} + \lambda(2\mathbf{i} + \mathbf{j} - 2\mathbf{k})$
 (d) $\mathbf{r} = 2\mathbf{i} - \mathbf{j} + 3\mathbf{k} + \lambda(-3\mathbf{i} + 2\mathbf{j})$

4 Find in the form $\mathbf{r}.\mathbf{n} = p$ an equation of the plane that passes through the point with position vector $\mathbf{a}$ and is perpendicular to the vector $\mathbf{n}$ where:
 (a) $\mathbf{a} = \mathbf{i} - \mathbf{j} + 2\mathbf{k}, \quad \mathbf{n} = 2\mathbf{i} + 4\mathbf{j} - \mathbf{k}$
 (b) $\mathbf{a} = 3\mathbf{i} - \mathbf{j} + 4\mathbf{k}, \quad \mathbf{n} = -\mathbf{i} + 3\mathbf{j} - 4\mathbf{k}$
 (c) $\mathbf{a} = 2\mathbf{i} + \mathbf{j} - \mathbf{k}, \quad \mathbf{n} = 2\mathbf{i} + 3\mathbf{j} - 4\mathbf{k}$
 (d) $\mathbf{a} = 2\mathbf{i} + \mathbf{j} + 2\mathbf{k}, \quad \mathbf{n} = -3\mathbf{i} + \mathbf{k}$
 (e) $\mathbf{a} = 4\mathbf{i} - \mathbf{j} + 3\mathbf{k}, \quad \mathbf{n} = 6\mathbf{i} + 4\mathbf{j} - 2\mathbf{k}$

5 Find a cartesian equation for each of the planes in question 4.

6 Verify that the point with position vector $\mathbf{a}$ lies in the given plane (λ, μ scalars) where:

(a) $\mathbf{a} = 2\mathbf{j} + \mathbf{k}$, $\quad \mathbf{r} = \mathbf{i} + 2\mathbf{j} + \mathbf{k} + \lambda(\mathbf{i} - \mathbf{j} + \mathbf{k}) + \mu(2\mathbf{i} - \mathbf{j} + \mathbf{k})$

(b) $\mathbf{a} = \mathbf{i} - \mathbf{j} + \mathbf{k}$, $\quad \mathbf{r} = \mathbf{i} + 3\mathbf{j} - \mathbf{k} + \lambda(-\mathbf{i} + \mathbf{j} - 2\mathbf{k}) + \mu(-\mathbf{i} - \mathbf{j} - \mathbf{k})$

(c) $\mathbf{a} = 24\mathbf{i} + 25\mathbf{j} - 9\mathbf{k}$, $\quad \mathbf{r} = 2\mathbf{i} - 3\mathbf{j} + \mathbf{k} + \lambda(7\mathbf{i} + 5\mathbf{j} - 2\mathbf{k}) + \mu(3\mathbf{i} - 4\mathbf{j} + \mathbf{k})$

7 Find (i) an equation in the form $\mathbf{r} = \mathbf{a} + \lambda\mathbf{b} + \mu\mathbf{c}$ (ii) a cartesian equation of the plane passing through the points:

(a) $(1, -1, 1)$, $(2, -4, 3)$, $(0, 1, -3)$

(b) $(4, 7, -1)$, $(1, 1, -4)$, $(2, -2, 3)$

(c) $(8, 1, -1)$, $(2, 6, -2)$, $(3, -3, 0)$

(d) $(2, 0, -3)$, $(1, 4, -1)$, $(2, -1, 0)$

8 Find a cartesian equation of the plane containing the points:

(a) $(1, 1, 1)$, $(2, 1, 0)$, $(2, 2, -1)$

(b) $(2, 1, -1)$, $(-2, -1, -5)$, $(0, -4, 3)$

(c) $(1, 1, 2)$, $(3, 4, 1)$, $(-5, 1, -1)$

(d) $(4, 0, 0)$, $(0, 3, 0)$, $\left(0, 0, -\frac{1}{2}\right)$

9 Find the coordinates of the point of intersection of the line l and the plane Π where

(a) l: $\mathbf{r} = \mathbf{i} - 2\mathbf{j} + \mathbf{k} + t(-\mathbf{i} + 2\mathbf{k})$, t scalar

Π: $\mathbf{r}.(2\mathbf{i} + \mathbf{j} + 2\mathbf{k}) = 4$

(b) l: $\mathbf{r} = 5\mathbf{i} - 2\mathbf{j} - 3\mathbf{k} + t(2\mathbf{i} - 3\mathbf{j} - 5\mathbf{k})$, t scalar

Π: $\mathbf{r}.(6\mathbf{i} + 2\mathbf{j} - 5\mathbf{k}) = 10$

10 Find an equation of the plane, in the form $\mathbf{r}.\mathbf{n} = p$, which contains the line l and the point with position vector $\mathbf{a}$ where

(a) l: $\mathbf{r} = t(2\mathbf{i} + 3\mathbf{j} - \mathbf{k})$, $\quad \mathbf{a} = \mathbf{i} + 4\mathbf{k}$

(b) l: $\mathbf{r} = 4\mathbf{i} + \mathbf{j} - 2\mathbf{k} + t(-\mathbf{i} + \mathbf{j} + 4\mathbf{k})$, $\quad \mathbf{a} = -\mathbf{i} + \mathbf{j} + 2\mathbf{k}$

(c) l: $\mathbf{r} = -2\mathbf{i} + 3\mathbf{j} - \mathbf{k} + t(2\mathbf{i} - \mathbf{j} - 3\mathbf{k})$, $\quad \mathbf{a} = -3\mathbf{i} + \mathbf{j} + 2\mathbf{k}$

11 Find a cartesian equation of the plane which passes through the origin O and contains the line with equations

$$\frac{x-1}{2} = \frac{y-2}{3} = \frac{z-3}{4}$$

12 Referred to an origin O, the points A, B, C have coordinates $(3, 2, 0)$, $(1, 0, 1)$, $(2, 2, 2)$ respectively.

(a) Find a cartesian equation of the plane ABC.

(b) Show that $D(4, 4, 1)$ lies in the plane.

(c) Show that AB and DC are parallel.

(d) Find the coordinates of the point where the lines AC and BD meet.

4.6 Distance of a point from a plane

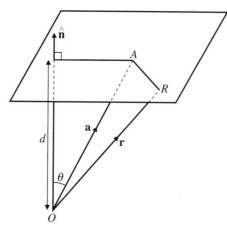

The diagram shows a plane which contains the point A, with position vector **a**, and also contains any other point R, with position vector **r**. The unit vector $\hat{\mathbf{n}}$ is perpendicular to the plane. The line OA makes an angle θ with $\hat{\mathbf{n}}$ and d is the distance of O from the plane.

Then:
$$d = a \cos \theta$$
$$= a \cdot 1 \cdot \cos \theta$$
$$= |\mathbf{a}| \, |\hat{\mathbf{n}}| \cos \theta, \text{ since } \hat{\mathbf{n}} \text{ is a unit vector}$$

So
$$d = \mathbf{a} \cdot \hat{\mathbf{n}}$$

Now an equation of the plane is $\mathbf{r} \cdot \mathbf{n} = \mathbf{a} \cdot \mathbf{n}$. Thus if you replace the vector **n** by $\hat{\mathbf{n}}$ in the scalar product form of the equation of the plane you get $\mathbf{r} \cdot \hat{\mathbf{n}} = d$ where d is the perpendicular distance of the origin from the plane.

If you have three parallel planes and if **n** is a vector perpendicular to one of these planes, then **n** will be perpendicular to each of the planes. Suppose that the equations of the planes are $\mathbf{r} \cdot \mathbf{n} = u$, $\mathbf{r} \cdot \mathbf{n} = v$ and $\mathbf{r} \cdot \mathbf{n} = -w$, then if you rewrite these equations as $\mathbf{r} \cdot \hat{\mathbf{n}} = a$, $\mathbf{r} \cdot \hat{\mathbf{n}} = b$ and $\mathbf{r} \cdot \hat{\mathbf{n}} = -c$ where $a = \dfrac{u}{|\mathbf{n}|}$, $b = \dfrac{v}{|\mathbf{n}|}$ and $c = \dfrac{w}{|\mathbf{n}|}$ then a, b and c are the perpendicular distances from O to each of the planes. The significance of the minus sign is that it indicates that the third plane is on the opposite side of the origin to the other two planes.

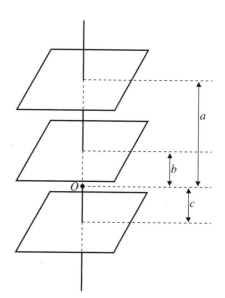

Example 13
Find the perpendicular distance of the origin from the plane with equation $\mathbf{r} \cdot (2\mathbf{i} + 4\mathbf{j} - 3\mathbf{k}) = 7$.

$2\mathbf{i} + 4\mathbf{j} - 3\mathbf{k}$ is perpendicular to the plane,
$$|2\mathbf{i} + 4\mathbf{j} - 3\mathbf{k}| = \sqrt{(4 + 16 + 9)} = \sqrt{29}$$

So a unit vector perpendicular to the plane is

$$\frac{2\mathbf{i} + 4\mathbf{j} - 3\mathbf{k}}{\sqrt{29}}$$

You can thus write the equation of the plane as

$$\frac{1}{\sqrt{29}}[\mathbf{r}.(2\mathbf{i} + 4\mathbf{j} - 3\mathbf{k})] = \frac{7}{\sqrt{29}}$$

The distance of O from the plane is $\dfrac{7}{\sqrt{29}}$.

Example 14

Find the distance of the point $(3, 1, 6)$ from the plane with equation $\mathbf{r}.(2\mathbf{i} - \mathbf{j} - 2\mathbf{k}) = 13$.

The equation of a plane is

$$\mathbf{r}.\mathbf{n} = \mathbf{a}.\mathbf{n}$$

where $\mathbf{a}$ is the position vector of a point in the plane. So an equation of the plane through $(3, 1, 6)$ parallel to $\mathbf{r}.(2\mathbf{i} - \mathbf{j} - 2\mathbf{k}) = 13$ is

$$\mathbf{r}.(2\mathbf{i} - \mathbf{j} - 2\mathbf{k}) = (3\mathbf{i} + \mathbf{j} + 6\mathbf{k}).(2\mathbf{i} - \mathbf{j} - 2\mathbf{k})$$

or: $\qquad\qquad \mathbf{r}.(2\mathbf{i} - \mathbf{j} - 2\mathbf{k}) = 6 - 1 - 12 = -7$

Now you can write $\mathbf{r}.(2\mathbf{i} - \mathbf{j} - 2\mathbf{k}) = -7$ as

$$\tfrac{1}{3}\mathbf{r}.(2\mathbf{i} - \mathbf{j} - 2\mathbf{k}) = -\tfrac{7}{3}$$

and you can write $\mathbf{r}.(2\mathbf{i} - \mathbf{j} - 2\mathbf{k}) = 13$ as

$$\tfrac{1}{3}\mathbf{r}.(2\mathbf{i} - \mathbf{j} - 2\mathbf{k}) = \tfrac{13}{3}$$

where $\tfrac{1}{3}(2\mathbf{i} - \mathbf{j} - \mathbf{k})$ is a unit vector perpendicular to each plane. You know that these two planes lie on opposite sides of the origin because $-\tfrac{7}{3}$ and $\tfrac{13}{3}$ have opposite signs. So the distance between the planes, and hence the distance of $(3, 1, 6)$ from $\mathbf{r}.(2\mathbf{i} - \mathbf{j} - 2\mathbf{k}) = 13$, is

$$\tfrac{7}{3} + \tfrac{13}{3} = \tfrac{20}{3} = 6\tfrac{2}{3}$$

4.7 The line of intersection of two planes

In general when two planes intersect their intersection is a straight line. If you can write the equation of each plane in cartesian form, then by solving the two equations simultaneously you should be able to obtain an equation of the line of intersection, as the next example shows.

Example 15

Find, in vector form, an equation of the line of intersection of the plane $\mathbf{r}.(\mathbf{i}+\mathbf{j}+\mathbf{k})=3$ with the plane $\mathbf{r}.(\mathbf{i}+2\mathbf{j}+3\mathbf{k})=5$.

$\mathbf{r}.(\mathbf{i}+\mathbf{j}+\mathbf{k})=3$ can be written:

$$x+y+z=3 \tag{1}$$

and $\mathbf{r}.(\mathbf{i}+2\mathbf{j}+3\mathbf{k})=5$ can be written:

$$x+2y+3z=5 \tag{2}$$

$(2)-(1)$ gives $\qquad y+2z=2$

So: $\qquad y=2-2z$

Substituting in (1) gives:

$$x+2-2z+z=3$$

$$\Rightarrow \quad x-z=1$$

or $\qquad x=1+z$

If you let $z=\lambda$, say, then

$$\frac{x-1}{1}=\frac{2-y}{2}=\frac{z}{1}(=\lambda)$$

which are equations of the line of intersection in cartesian form.

So: $\qquad x=1+\lambda,\, y=2-2\lambda,\, z=\lambda$

If $\qquad \mathbf{r}=x\mathbf{i}+y\mathbf{j}+z\mathbf{k}$

then: $\qquad \mathbf{r}=(1+\lambda)\mathbf{i}+(2-2\lambda)\mathbf{j}+\lambda\mathbf{k}$

That is: $\qquad \mathbf{r}=\mathbf{i}+2\mathbf{j}+\lambda(\mathbf{i}-2\mathbf{j}+\mathbf{k})$

which is an equation of the line of intersection in vector form.

4.8 The angle between a line and a plane

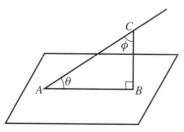

The angle between a line and a plane is the angle between the line and its projection on the plane. In the diagram, AC is the line and AB is its projection on the plane. So you have to find the angle $CAB=\theta$.

If you draw BC so that $\angle ABC = 90°$ then

$$\theta + \phi = 90°, \quad \text{so } \theta = 90° - \phi$$

and $$\sin \theta = \sin(90° - \phi) = \cos \phi$$

But if $\mathbf{n}$ is a vector perpendicular to the plane and $\mathbf{b}$ is a direction vector of the line (Book C4, page 64) then you know that

$$\mathbf{b.n} = |\mathbf{b}| \, |\mathbf{n}| \cos \phi$$

or $$\cos \phi = \left| \frac{\mathbf{b.n}}{|\mathbf{b}| \, |\mathbf{n}|} \right|$$

where you take the modulus to ensure that ϕ, and hence θ, is acute.

But $$\cos \phi = \sin \theta$$

so: $$\sin \theta = \left| \frac{\mathbf{b.n}}{|\mathbf{b}| \, |\mathbf{n}|} \right|$$

and hence:

■ $$\theta = \arcsin \left| \frac{\mathbf{b.n}}{|\mathbf{b}| \, |\mathbf{n}|} \right|$$

Example 16

Find the acute angle between the line with equations

$$\frac{x+1}{2} = y - 2 = \frac{z-3}{-2} = \lambda$$

and the plane with equation

$$2x + 3y - 7z = 5$$

If $$\frac{x+1}{2} = y - 2 = \frac{z-3}{-2} = \lambda$$

then: $$x = 2\lambda - 1, \quad y = \lambda + 2, \quad z = 3 - 2\lambda$$

Since $\mathbf{r} = x\mathbf{i} + y\mathbf{j} + z\mathbf{k}$, then a vector equation of the line is

$$\mathbf{r} = (2\lambda - 1)\mathbf{i} + (\lambda + 2)\mathbf{j} + (3 - 2\lambda)\mathbf{k}$$

That is: $$\mathbf{r} = -\mathbf{i} + 2\mathbf{j} + 3\mathbf{k} + \lambda(2\mathbf{i} + \mathbf{j} - 2\mathbf{k})$$

A vector equation of the plane is

$$\mathbf{r}.(2\mathbf{i} + 3\mathbf{j} - 7\mathbf{k}) = 5$$

If AC is part of the line and AB is its projection on the plane then you know that AC is parallel to $2\mathbf{i} + \mathbf{j} - 2\mathbf{k}$ and BC is parallel to $2\mathbf{i} + 3\mathbf{j} - 7\mathbf{k}$

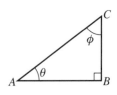

So: $$(2\mathbf{i} + \mathbf{j} - 2\mathbf{k}).(2\mathbf{i} + 3\mathbf{j} - 7\mathbf{k}) = \sqrt{9}\sqrt{62} \cos \phi$$

$$\cos \phi = \frac{4 + 3 + 14}{3\sqrt{62}}$$

Thus
$$\sin \theta = \frac{21}{3\sqrt{62}} = \frac{7}{\sqrt{62}}$$

and
$$\theta = 63° \text{ (nearest degree)}$$

4.9 Angle between two planes

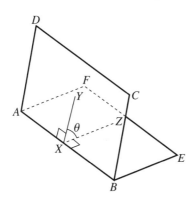

The angle between two planes is the angle between two perpendiculars (one in each plane) drawn from a point on their line of intersection. If you are given vector equations of the two planes you need to be able to calculate $\angle YXZ = \theta$. So from Y you draw the perpendicular to the plane $ABCD$ and from Z you draw the perpendicular to the plane $ABEF$.

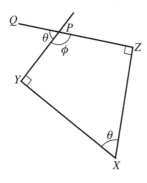

If the two perpendiculars meet at P then

$$\angle YXZ + \angle YPZ = 180°$$

(since the angle sum of a quadrilateral is $360°$ and $\angle PYX = \angle PZX = 90°$).

That is:
$$\theta + \phi = 180°$$

and hence
$$\angle QPY = \angle YXZ$$

The acute angle θ between the planes is therefore equal to the acute angle between the normals to the two planes.

If $\mathbf{n}_1$ is a vector perpendicular to plane $ABCD$ and if $\mathbf{n}_2$ is a vector perpendicular to plane $ABEF$ then

$$\mathbf{n}_1.\mathbf{n}_2 = |\mathbf{n}_1|\,|\mathbf{n}_2|\cos\theta$$

and so

$$\cos\theta = \left|\frac{\mathbf{n}_1.\mathbf{n}_2}{|\mathbf{n}_1|\,|\mathbf{n}_2|}\right|$$

where you take the modulus of the answer because you need the angle to be acute.

Example 17

Find the angle between the planes with equations

$$\mathbf{r}.(2\mathbf{i}+\mathbf{j}+3\mathbf{k}) = 5$$

and

$$\mathbf{r}.(2\mathbf{i}+3\mathbf{j}+\mathbf{k}) = 7$$

A vector normal to the first plane is $2\mathbf{i}+\mathbf{j}+3\mathbf{k}$. A vector normal to the second plane is $2\mathbf{i}+3\mathbf{j}+\mathbf{k}$. So if θ is the angle between the planes then

$$(2\mathbf{i}+\mathbf{j}+3\mathbf{k}).(2\mathbf{i}+3\mathbf{j}+\mathbf{k}) = \sqrt{14}\sqrt{14}\cos\theta$$

$$\cos\theta = \frac{4+3+3}{14} = \frac{10}{14} = \frac{5}{7}$$

$$\theta = 44° \text{ (nearest degree)}$$

4.10 Shortest distance between two skew lines

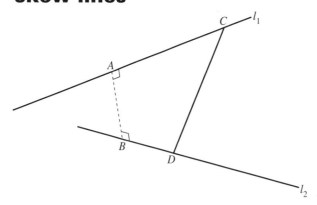

If l_1 and l_2 are two skew lines with equations

$$\mathbf{r} = \mathbf{a} + \lambda\mathbf{b}$$

and

$$\mathbf{r} = \mathbf{c} + \mu\mathbf{d}$$

respectively, where λ, μ are scalars, then the shortest distance between the lines is the distance AB where AB is perpendicular to both l_1 and l_2. Now **b** is the direction vector of l_1 and **d** is the direction vector of l_2. So **b** × **d** is a vector in the direction of AB.

Consequently $\dfrac{\mathbf{b} \times \mathbf{d}}{|\mathbf{b} \times \mathbf{d}|}$ is a unit vector in the direction AB.

Now the position vector of an arbitrary point C on l_1 is $\mathbf{a} + \lambda\mathbf{d}$ and the position vector of an arbitrary point D on l_2 is $\mathbf{c} + \mu\mathbf{d}$.

So the vector $\overrightarrow{DC}$ is

$$\mathbf{a} + \lambda\mathbf{d} - (\mathbf{c} + \mu\mathbf{d})$$

$$= \mathbf{a} - \mathbf{c} + \lambda\mathbf{b} - \mu\mathbf{d}$$

The distance that we want is the projection of this vector in the direction of AB.

So you require:

$$[\mathbf{a} - \mathbf{c} + \lambda\mathbf{b} - \mu\mathbf{d}] \, . \, \frac{\mathbf{b} \times \mathbf{d}}{|\mathbf{b} \times \mathbf{d}|}$$

$$= \frac{(\mathbf{a} - \mathbf{c}) \, . \, \mathbf{b} \times \mathbf{d}}{|\mathbf{b} \times \mathbf{d}|} + \frac{\lambda\mathbf{b} \, . \, \mathbf{b} \times \mathbf{d}}{|\mathbf{b} \times \mathbf{d}|} - \frac{\mu\mathbf{d} \, . \, \mathbf{b} \times \mathbf{d}}{|\mathbf{b} \times \mathbf{d}|}$$

By definition, the vector **b** × **d** is perpendicular to both **b** and **d**.

So: $$\mathbf{b} \, . \, \mathbf{b} \times \mathbf{d} = 0$$

and $$\mathbf{d} \, . \, \mathbf{b} \times \mathbf{d} = 0$$

This means that

$$\frac{\lambda\mathbf{b} \, . \, \mathbf{b} \times \mathbf{d}}{|\mathbf{b} \times \mathbf{d}|} = 0$$

and $$\frac{\mu\mathbf{b} \, . \, \mathbf{b} \times \mathbf{d}}{|\mathbf{b} \times \mathbf{d}|} = 0$$

So the shortest distance between the skew lines is given by

$$\frac{(\mathbf{a} - \mathbf{c}) \, . \, \mathbf{b} \times \mathbf{d}}{|\mathbf{b} \times \mathbf{d}|}$$

Well almost! When you found **b** × **d** you might have evaluated the negative direction rather than the positive direction and so your answer might come out to be negative.

Since this is meaningless for a distance, you need, finally, to take the modulus.

■ **The shortest distance between two skew lines with equations**

$$\mathbf{r} = \mathbf{a} + \lambda\mathbf{b}$$

and $$\mathbf{r} = \mathbf{c} + \mu\mathbf{d}$$

where λ, μ are scalars is

$$\left| \frac{(\mathbf{a} - \mathbf{c}) \, . \, \mathbf{b} \times \mathbf{d}}{|\mathbf{b} \times \mathbf{d}|} \right|$$

Example 18

Find the shortest distance between the lines l_1 and l_2 with equations

$$\mathbf{r} = 2\mathbf{i} + 3\mathbf{j} - 5\mathbf{k} + \lambda(3\mathbf{i} + 2\mathbf{j} - 4\mathbf{k})$$

and
$$\mathbf{r} = 3\mathbf{i} - \mathbf{j} + 4\mathbf{k} + \mu(5\mathbf{i} - 2\mathbf{j} - 4\mathbf{k})$$

respectively.

A vector in the direction of the line of shortest distance is

$$(3\mathbf{i} + 2\mathbf{j} - 4\mathbf{k}) \times (5\mathbf{i} - 2\mathbf{j} - 4\mathbf{k})$$

$$= \begin{vmatrix} \mathbf{i} & \mathbf{j} & \mathbf{k} \\ 3 & 2 & -4 \\ 5 & -2 & -4 \end{vmatrix}$$

$$= \mathbf{i}(-8-8) - \mathbf{j}(-12+20) + \mathbf{k}(-6-10)$$

$$= -16\mathbf{i} - 8\mathbf{j} - 16\mathbf{k}$$

So a unit vector in this direction is:

$$\frac{-16\mathbf{i} - 8\mathbf{j} - 16\mathbf{k}}{\sqrt{16^2 + 8^2 + 16^2}}$$

$$= \frac{-16\mathbf{i} - 8\mathbf{j} - 16\mathbf{k}}{8\sqrt{4 + 1 + 4}}$$

$$= \frac{-2\mathbf{i} - \mathbf{j} - 2\mathbf{k}}{3}$$

A vector joining an arbitrary point on l_1 to an arbitrary point on l_2 is:

$$2\mathbf{i} + 3\mathbf{j} - 5\mathbf{k} - 3\mathbf{i} + \mathbf{j} - 4\mathbf{k} + \lambda(3\mathbf{i} + 2\mathbf{j} - 4\mathbf{k}) - \mu(5\mathbf{i} - 2\mathbf{j} - 4\mathbf{k})$$

$$= -\mathbf{i} + 4\mathbf{j} - 9\mathbf{k} + \lambda(3\mathbf{i} + 2\mathbf{j} - 4\mathbf{k}) - \mu(5\mathbf{i} - 2\mathbf{j} - 4\mathbf{k})$$

The shortest distance is the length of the projection of this vector in the direction $\dfrac{-2\mathbf{i} - \mathbf{j} - 2\mathbf{k}}{3}$. That is:

$$\frac{[-\mathbf{i} + 4\mathbf{j} - 9\mathbf{k} + \lambda(3\mathbf{i} + 2\mathbf{j} - 4\mathbf{k}) - \mu(5\mathbf{i} - 2\mathbf{j} - 4\mathbf{k})].(-2\mathbf{i} - \mathbf{j} - 2\mathbf{k})}{3}$$

Now $-2\mathbf{i} - \mathbf{j} - 2\mathbf{k}$ is perpendicular to both $3\mathbf{i} + 2\mathbf{j} - 4\mathbf{k}$ and $5\mathbf{i} - 2\mathbf{j} - 4\mathbf{k}$.

So: $$(3\mathbf{i} + 2\mathbf{j} - 4\mathbf{k}).(-2\mathbf{i} - \mathbf{j} - 2\mathbf{k}) = 0$$

and $$(5\mathbf{i} - 2\mathbf{j} - 4\mathbf{k}).(-2\mathbf{i} - \mathbf{j} - 2\mathbf{k}) = 0$$

So the shortest distance is

$$\left| \frac{(-\mathbf{i} + 4\mathbf{j} - 9\mathbf{k}).(-2\mathbf{i} - \mathbf{j} - 2\mathbf{k})}{3} \right|$$

$$= \frac{2 - 4 + 18}{3} = \frac{16}{3}$$

Exercise 4D

1 Find the distance from the origin to the plane with equation:
   (a) $\mathbf{r}.(2\mathbf{i}+\mathbf{j}-3\mathbf{k})=5$
   (b) $\mathbf{r}.(-3\mathbf{i}+\mathbf{j}+6\mathbf{k})=24$
   (c) $\mathbf{r}.(5\mathbf{i}-10\mathbf{j}+4\mathbf{k})=17$
   (d) $\mathbf{r}.(3\mathbf{i}+12\mathbf{j}-4\mathbf{k})=62$
   (e) $x-y+2z=15$

2 Find the distance from the given point to the given plane:
   (a) $(1,2,-3),\quad \mathbf{r}.(\mathbf{i}+2\mathbf{j}-\mathbf{k})=23$
   (b) $(1,-3,2),\quad \mathbf{r}.(2\mathbf{i}-\mathbf{j}+3\mathbf{k})=46$
   (c) $(4,1,-7),\quad 2x+6y-3z=14$
   (d) $(1,-3,5),\quad \mathbf{r}.(4\mathbf{i}-2\mathbf{j}-4\mathbf{k})=10$
   (e) $(4,-8,-1),\quad 4x+y-7z=42$

3 Find the position vector of the point where the line with equation $\mathbf{r}=\mathbf{i}+\mathbf{j}-2\mathbf{k}+\lambda(\mathbf{i}+5\mathbf{k})$ cuts the plane with equation $\mathbf{r}.(2\mathbf{i}+\mathbf{j}-\mathbf{k})=2$.

4 Find, in degrees to $0.1°$, the acute angle between the given line and plane:
   (a) $\mathbf{r}=\mathbf{i}-2\mathbf{k}+\lambda(\mathbf{i}+\mathbf{j}-3\mathbf{k})$ and $\mathbf{r}.(2\mathbf{i}-\mathbf{j}+4\mathbf{k})=10$
   (b) $\mathbf{r}=3\mathbf{i}+\mathbf{j}-\mathbf{k}+\lambda(4\mathbf{i}-7\mathbf{k})$ and $\mathbf{r}.(-\mathbf{i}+4\mathbf{j}-6\mathbf{k})=24$
   (c) $\mathbf{r}=-6\mathbf{i}+2\mathbf{j}-4\mathbf{k}+\lambda(-2\mathbf{i}+3\mathbf{j}-6\mathbf{k})$ and $\mathbf{r}.(2\mathbf{i}-6\mathbf{j}+5\mathbf{k})=63$

5 Find, in degrees to $0.1°$, the angle between the planes with equations:
   (a) $\mathbf{r}.(\mathbf{i}+2\mathbf{j}-2\mathbf{k})=3$ and $\mathbf{r}.(2\mathbf{i}+2\mathbf{j}-\mathbf{k})=6$
   (b) $\mathbf{r}.(\mathbf{i}-\mathbf{j}-\sqrt{2}\mathbf{k})=5$ and $\mathbf{r}.(7\mathbf{j}+\mathbf{k})=10$
   (c) $x-2y-5z=7$ and $3x+7y-z=4$

6 Find the distance from the origin to the plane with equations $\mathbf{r}.(2\mathbf{i}-2\mathbf{j}+\mathbf{k})=6$.

7 Find the cosine of the acute angle between the planes with equations $2x+3y-4z=5$ and $6x-2y-3z=4$.

8 Find a vector equation of the line of intersection of the planes with equations:
   (a) $\mathbf{r}.(2\mathbf{i}+\mathbf{j}-3\mathbf{k})=28$ and $\mathbf{r}.(4\mathbf{i}-7\mathbf{j}+\mathbf{k})=31$
   (b) $x+6y+z=-10,\quad 3x+2y-z=-1$
   (c) $\mathbf{r}.(\mathbf{i}+2\mathbf{j}+\mathbf{k})=-5$ and $\mathbf{r}.(2\mathbf{i}-3\mathbf{j}+\mathbf{k})=15$

9   Find, in degrees to 0.1°, the acute angle between the planes with equations $3x + 4y + 2z = 7$ and $2x - 3y + z = 9$.

10   Find the distance of the origin from the plane with equation $\mathbf{r}.(3\mathbf{i} - 4\mathbf{j} - 12\mathbf{k}) = 26$.

11   (a) Find, in cartesian form, the equation of the plane P which passes through the origin and contains the line with equations

$$\frac{x - 1}{2} = \frac{y - 2}{3} = \frac{z - 3}{4}$$

(b) Find, in degrees to 0.1°, the acute angle between P and the plane with equation

$$4x + y - z = 3$$

12   A line has equation
$$\mathbf{r} = 3\mathbf{i} - 5\mathbf{j} + 2\mathbf{k} + \lambda(2\mathbf{i} - 4\mathbf{j} + \mathbf{k})$$
and a plane has equation
$$\mathbf{r}.(3\mathbf{i} - \mathbf{j} - 5\mathbf{k}) = 1$$
Find the acute angle between the line and the plane.

13   The planes with equations
$$2x - y + 3z + 3 = 0 \quad \text{and} \quad x + 10y = 21$$
meet in a line L.
The planes with equations
$$2x - y = 0 \quad \text{and} \quad 7x + z = 6$$
meet in a line M.
Show that L and M meet at a point. Show further that L and M both lie in the plane with equation $x + 3y + z = 6$.

14   Referred to a fixed origin O, the lines l_1 and l_2 have equations
$$\mathbf{r} = 3\mathbf{i} + 6\mathbf{j} + \mathbf{k} + s(2\mathbf{i} + 3\mathbf{j} - \mathbf{k})$$
and $\qquad\qquad \mathbf{r} = 3\mathbf{i} - \mathbf{j} + 4\mathbf{k} + t(\mathbf{i} - 2\mathbf{j} + \mathbf{k})$
respectively, where s and t are scalar parameters.
(a) Show that l_1 and l_2 intersect and determine the position vector of their point of intersection.
(b) Show that the vector $-\mathbf{i} + 3\mathbf{j} + 7\mathbf{k}$ is perpendicular to both l_1 and l_2.
(c) Find, in the form $\mathbf{r}.\mathbf{n} = p$, an equation of the plane containing l_1 and l_2.   [E]

15 With respect to the origin O the points A, B, C have position vectors

$$5\mathbf{i} - \mathbf{j} - 3\mathbf{k}, \quad -4\mathbf{i} + 4\mathbf{j} - \mathbf{k}, \quad 5\mathbf{i} - 2\mathbf{j} + 11\mathbf{k}$$

respectively. Find

(a) a vector equation for the line BC

(b) a vector equation for the plane OAB

(c) the cosine of the acute angle between the lines OA and OB.

Obtain, in the form $\mathbf{r}.\mathbf{n} = p$, a vector equation for P, the plane which passes through A and is perpendicular to BC.

Find cartesian equations for

(d) the plane P

(e) the line BC. [E]

16 Show that

$$\mathbf{r} = \mathbf{a} + s(\mathbf{b} - \mathbf{a}) + t(\mathbf{c} - \mathbf{a})$$

is an equation of the plane which passes through the non-collinear points whose position vectors are $\mathbf{a}$, $\mathbf{b}$, $\mathbf{c}$, where $\mathbf{r}$ is the position vector of a general point on the plane and s and t are scalars. Find a cartesian equation of the plane containing the points $(1, 1, -1)$, $(2, 0, 1)$ and $(3, 2, 1)$, and show that the points $(2, 1, 2)$ and $(0, -2, -2)$ are equidistant from, and on opposite sides of, this plane. [E]

17 A plane passes through the three points A, B, C whose position vectors, referred to an origin O, are $(\mathbf{i} + 3\mathbf{j} + 3\mathbf{k})$, $(3\mathbf{i} + \mathbf{j} + 4\mathbf{k})$, $(2\mathbf{i} + 4\mathbf{j} + \mathbf{k})$ respectively. Find, in the form $(l\mathbf{i} + m\mathbf{j} + n\mathbf{k})$, a unit vector normal to this plane. Find also a cartesian equation of the plane, and the perpendicular distance from the origin to this plane. [E]

18 Show that the vector $\mathbf{i} + \mathbf{k}$ is perpendicular to the plane with vector equation

$$\mathbf{r} = \mathbf{i} + s\mathbf{j} + t(\mathbf{i} - \mathbf{k})$$

Find the perpendicular distance from the origin to this plane. Hence, or otherwise, obtain a cartesian equation of the plane. [E]

19 Three planes have equations

$$x - 6y - z = 5$$
$$3x + 2y + z = -1$$
$$5x + pz = q$$

Show that

(a) the planes have a common point of intersection unless $p = 1$

(b) when $p = 1$, $q = 2$, the planes intersect in pairs in three parallel lines

(c) when $p = 1$, $q = 1$, the planes have a common line of intersection.

Give equations for the line of intersection in (c). [E]

20 Show that the lines l_1, l_2, with vector equations
$$\mathbf{r} = 5\mathbf{i} - 2\mathbf{j} + 3\mathbf{k} + \lambda(-3\mathbf{i} + \mathbf{j} - \mathbf{k})$$
$$\mathbf{r} = 10\mathbf{i} - 3\mathbf{j} + 6\mathbf{k} + \mu(4\mathbf{i} - \mathbf{j} + 2\mathbf{k})$$
respectively, intersect and find a vector equation of the plane Π containing l_1 and l_2.

Show that the point Q with position vector $(6\mathbf{i} + 7\mathbf{j} - 2\mathbf{k})$ lies on the line which is perpendicular to Π and which passes through the intersection of l_1 and l_2. Find a vector equation of the plane which passes through Q and is parallel to Π. [E]

21 With respect to a fixed origin O, the straight lines l_1 and l_2 are given by
$$l_1 : \mathbf{r} = \mathbf{i} - \mathbf{j} + \lambda(2\mathbf{i} + \mathbf{j} - 2\mathbf{k})$$
$$l_2 : \mathbf{r} = \mathbf{i} + 2\mathbf{j} + 2\mathbf{k} + \mu(-3\mathbf{i} + 4\mathbf{k})$$
where λ and μ are scalar parameters.

(a) Show that the lines intersect.

(b) Find the position vector of their point of intersection.

(c) Find the cosine of the acute angle contained between the lines.

(d) Find a vector equation of the plane containing the lines. [E]

22 The position vectors of the points A, B, C are $\mathbf{a}$, $\mathbf{b}$ and $\mathbf{c}$ respectively, where
$$\mathbf{a} = -2\mathbf{i} + \mathbf{j}, \quad \mathbf{b} = \mathbf{i} + 2\mathbf{j} - 2\mathbf{k}, \quad \mathbf{c} = 5\mathbf{j} - 4\mathbf{k}$$

(a) Find $(\mathbf{b} - \mathbf{a}) \times (\mathbf{c} - \mathbf{a})$ and hence, or otherwise, find an equation of the plane ABC in the form $\mathbf{r}.\mathbf{n} = p$ and the area of the triangle ABC.

(b) Find a vector equation of the plane which passes through A and which is perpendicular to both the plane ABC and the plane with equation $(\mathbf{r} - \mathbf{a}).\mathbf{b} = 0$.

(c) Find cartesian equations for the line BC. [E]

23 Planes Π_1 and Π_2 have equations given by

$$\Pi_1 : \mathbf{r}.(2\mathbf{i} - \mathbf{j} + \mathbf{k}) = 0$$
$$\Pi_2 : \mathbf{r}.(\mathbf{i} + 5\mathbf{j} + 3\mathbf{k}) = 1$$

(a) Show that the point $A(2, -2, 3)$ lies in Π_2.

(b) Show that Π_1 is perpendicular to Π_2.

(c) Find, in vector form, an equation of the straight line through A which is perpendicular to Π_1.

(d) Determine the coordinates of the point where this line meets Π_1.

(e) Find the perpendicular distance of A from Π_1.

(f) Find a vector equation of the plane through A parallel to Π_1. [E]

24 Show that the line with equations

$$\frac{x - 4}{1} = \frac{y - 5}{2} = \frac{z - 6}{3}$$

and the line with equations

$$\frac{x - 1}{4} = \frac{y - 2}{5} = \frac{z - 3}{6}$$

intersect. Find an equation for the plane in which they lie and the coordinates of their point of intersection. [E]

25 The coordinates of the four points A, B, C, D are respectively $(1, 2, 1)$, $(-1, 0, 2)$, $(2, 1, 3)$ and $(3, -1, 1)$. Find the shortest distance between the lines AB and CD.

26 Show that the shortest distance between the line with equations

$$\frac{x + 4}{3} = \frac{y - 3}{2} = \frac{z + 6}{5}$$

and the line with equations $x - 2y - z = 0$, $x - 10y - 3z = -7$ is $\frac{1}{2}\sqrt{14}$.

27 Find the shortest distance between the lines with vector equations

$$\mathbf{r} = 3\mathbf{i} + s\mathbf{j} - \mathbf{k}$$

and

$$\mathbf{r} = 9\mathbf{i} - 2\mathbf{j} - \mathbf{k} + t(\mathbf{i} - 2\mathbf{j} + \mathbf{k})$$

where s, t are scalars. [E]

28 Find the shortest distance between the lines PQ and RS where P, Q, R, S have coordinates $(2, 1, 3)$, $(1, 2, 1)$, $(-1, -2, -2)$ and $(1, -4, 0)$ respectively.

29 Obtain the shortest distance between the lines with equations

$$\mathbf{r} = (3s - 3)\mathbf{i} - s\mathbf{j} + (s + 1)\mathbf{k}$$

and $$\mathbf{r} = (3 + t)\mathbf{i} + (2t - 2)\mathbf{j} + \mathbf{k}$$

where s, t are parameters. [E]

SUMMARY OF KEY POINTS

1 The vector product of $\mathbf{a}$ and $\mathbf{b}$ is

$$\mathbf{a} \times \mathbf{b} = |\mathbf{a}|\,|\mathbf{b}|\,\sin\theta\,\hat{\mathbf{n}}$$

where θ is the angle between $\mathbf{a}$ and $\mathbf{b}$ and $\hat{\mathbf{n}}$ is a unit vector perpendicular to both $\mathbf{a}$ and $\mathbf{b}$ which is in the direction that a right-handed corkscrew would move when turned from $\mathbf{a}$ to $\mathbf{b}$.

2 $\mathbf{a} \times \mathbf{b} = -\mathbf{b} \times \mathbf{a}$

3 (i) $\mathbf{i} \times \mathbf{j} = \mathbf{k}$
(ii) $\mathbf{j} \times \mathbf{k} = \mathbf{i}$
(iii) $\mathbf{k} \times \mathbf{i} = \mathbf{j}$

where $\mathbf{i}$, $\mathbf{j}$, $\mathbf{k}$ are the unit vectors in the directions of the positive x-, y- and z-axes respectively.

4 (i) $\mathbf{i} \times \mathbf{i} = \mathbf{0}$
(ii) $\mathbf{j} \times \mathbf{j} = \mathbf{0}$
(iii) $\mathbf{k} \times \mathbf{k} = \mathbf{0}$

where $\mathbf{i}$, $\mathbf{j}$, $\mathbf{k}$ are the unit vectors in the directions of the positive x-, y- and z-axes respectively.

5 If $\qquad\qquad \mathbf{a} = a_1\mathbf{i} + a_2\mathbf{j} + a_3\mathbf{k}$
and $\qquad\qquad \mathbf{b} = b_1\mathbf{i} + b_2\mathbf{j} + b_3\mathbf{k}$

then $\qquad \mathbf{a} \times \mathbf{b} = \begin{vmatrix} \mathbf{i} & \mathbf{j} & \mathbf{k} \\ a_1 & a_2 & a_3 \\ b_1 & b_2 & b_3 \end{vmatrix}$

6 If $\mathbf{a} \times \mathbf{b} = \mathbf{0}$ then either $\mathbf{a} = \mathbf{0}$ or $\mathbf{b} = \mathbf{0}$ or $\mathbf{a}$ and $\mathbf{b}$ are parallel.

7
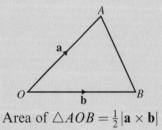

Area of $\triangle AOB = \frac{1}{2}|\mathbf{a} \times \mathbf{b}|$

8

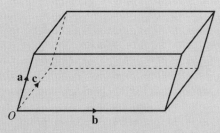

The volume of the parallelepiped is

$$\mathbf{a}.\mathbf{b} \times \mathbf{c}$$

9 If $\mathbf{a} = a_1\mathbf{i} + a_2\mathbf{j} + a_3\mathbf{k}$
$\mathbf{b} = b_1\mathbf{i} + b_2\mathbf{j} + b_3\mathbf{k}$
$\mathbf{c} = c_1\mathbf{i} + c_2\mathbf{j} + c_3\mathbf{k}$

then $\mathbf{a}.\mathbf{b} \times \mathbf{c} = \begin{vmatrix} a_1 & a_2 & a_3 \\ b_1 & b_2 & b_3 \\ c_1 & c_2 & c_3 \end{vmatrix}$

10

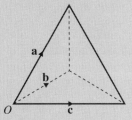

The volume of the tetrahedron is

$$\tfrac{1}{6}\mathbf{a}.\mathbf{b} \times \mathbf{c}$$

11 The equation of the line passing through A, with position vector $\mathbf{a}$, and the point R, with position vector $\mathbf{r}$, and which is parallel to the vector $\mathbf{b}$, is

$$(\mathbf{r} - \mathbf{a}) \times \mathbf{b} = \mathbf{0}$$

12 The equation of the plane containing the points A and R, with position vectors $\mathbf{a}$ and $\mathbf{r}$ respectively, is $\mathbf{r}.\mathbf{n} = p$, where $p = \mathbf{a}.\mathbf{n}$ and $\mathbf{n}$ is a vector perpendicular to the plane.

13 The vector equation of a plane passing through the point with position vector $\mathbf{a}$ and where $\mathbf{b}$ and $\mathbf{c}$ are non-parallel vectors in the plane, neither of which is zero is

$$\mathbf{r} = \mathbf{a} + \lambda\mathbf{b} + \mu\mathbf{c},$$

where λ and μ are scalars.

14 The distance d from the origin to the plane containing the point with position vector $\mathbf{r}$ is

$$d = \mathbf{r}.\hat{\mathbf{n}}$$

where $\hat{\mathbf{n}}$ is a unit vector perpendicular to the plane.

15 The acute angle θ between a line, with direction vector **b**, and a plane is

$$\arcsin \left| \frac{\mathbf{b} \cdot \mathbf{n}}{|\mathbf{b}| \, |\mathbf{n}|} \right|$$

where **n** is a vector perpendicular to the plane.

16 The acute angle θ between two planes is given by

$$\cos \theta = \left| \frac{\mathbf{n}_1 \cdot \mathbf{n}_2}{|\mathbf{n}_1| \, |\mathbf{n}_2|} \right|$$

where $\mathbf{n}_1$ is a vector perpendicular to one of the planes and $\mathbf{n}_2$ is a vector perpendicular to the other plane.

17 The shortest distance between the lines with equations $\mathbf{r} = \mathbf{a} + \lambda \mathbf{b}$ and $\mathbf{r} = \mathbf{c} + \mu \mathbf{d}$ where λ, μ are scalars is given by

$$\left| \frac{(\mathbf{a} - \mathbf{c}) \cdot \mathbf{b} \times \mathbf{d}}{|\mathbf{b} \times \mathbf{d}|} \right|$$

Numerical methods

5.1 Step-by-step methods for solving differential equations

There are many first order and second order differential equations that cannot be solved analytically. Suppose that you have a differential equation of the type $\dfrac{dy}{dx} = g(x, y)$, where g is a function of the variables x and y and you know further that $y = y_0$ at $x = x_0$. In other words, you know that the point (x_0, y_0) is on the solution curve of the differential equation. If you take small steps of size h along the x-axis from where $x = x_0$ you get a set of points x_1, x_2, $x_3 \ldots$ where $x = x_1 = x_0 + h$, $x = x_2 = x_1 + h$, $x = x_3 = x_2 + h$, and so on.

In a step-by-step approach, you need a method which enables you to estimate successive values y_1, y_2, $y_3, \ldots$ so that the points (x_1, y_1), (x_2, y_2), $(x_3, y_3) \ldots$ lie approximately on the solution curve of the differential equation $\dfrac{dy}{dx} = g(x, y)$, where $y = y_0$ at $x = x_0$. Clearly, if you have a different starting point from (x_0, y_0), then you will build up a different approximate solution curve. So the question arises: how do you make the estimates y_1, y_2, $y_3 \ldots$? The solution depends on Taylor's series. From section 1.4 you have for $f(x + a)$, the Taylor series

$$f(x + a) = f(a) + xf'(a) + \frac{x^2}{2!}f''(a) + \ldots$$

Take $x = h$ and $a = x_0$ and you have

$$f(x_0 + h) = f(x_0) + hf'(x_0) + \frac{h^2}{2!}f''(x_0) + \ldots$$

Assume that h is small, so that h^2, $h^3 \ldots$ are negligible compared to h, then you have

$$f(x_0 + h) \approx f(x_0) + hf'(x_0)$$

Take $y_0 = f(x_0)$, $y_1 = f(x_0 + h)$ and $f'(x_0) = \left(\dfrac{dy}{dx}\right)_0$ and you have

$y_1 \approx y_0 + h\left(\dfrac{dy}{dx}\right)_0$ which gives on rearranging

■
$$\left(\frac{dy}{dx}\right)_0 \approx \frac{y_1 - y_0}{h}$$

Since also $\left(\frac{dy}{dx}\right)_0 = g(x_0, y_0)$, you have

$$y_1 \approx y_0 + h[g(x_0, y_0)]$$

as the estimated value of y at $x = x_1$. Then, by repeating the process, you can obtain

$$y_2 \approx y_1 + h[g(x_1, y_1)]$$

and so on. In general,

■
$$\boldsymbol{y_{n+1} \approx y_n + hg(x_n, y_n)}$$

Notice that the formula $\left(\frac{dy}{dx}\right)_0 \approx \frac{y_1 - y_0}{h}$, often called **Euler's method**, can also be obtained by geometrical considerations.

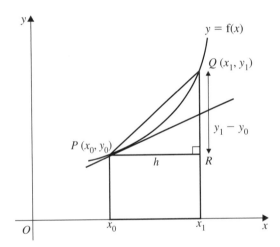

The point $P(x_0, y_0)$ lies on the curve with equation $y = f(x)$. The point $Q(x_1, y_1)$ also lies on the curve. The tangent to the curve at $P(x_0, y_0)$ has gradient $f'(x_0)$ which can be written as $\left(\frac{dy}{dx}\right)_0$. Now if you assume that $x_1 - x_0 = h$ and that h is small, then the gradient of the chord PQ is approximately equal to the gradient of the tangent to the curve at P. But since $QR = y_1 - y_0$, the gradient of the chord PQ is $\frac{y_1 - y_0}{h}$.

Hence:
$$\left(\frac{dy}{dx}\right)_0 \approx \frac{y_1 - y_0}{h}$$

Example 1

$$\frac{dy}{dx} = -xy - 2 \text{ and } y = 1 \text{ at } x = 0$$

Using a step length h of 0.1, find an estimate of y at $x = 0.3$, giving 4 decimal places in your answer.

Step 1
In the standard notation, you have $h = 0.1$, $y_0 = 1$ and $x_0 = 0$, and using the formula

$$y_1 \approx h\left[\left(\frac{dy}{dx}\right)_0\right] + y_0 = h(-x_0 y_0 - 2) + y_0$$

That is: $y_1 \approx 0.1(0 - 2) + 1 = 0.8$

Step 2
Now take $y_1 = 0.8$ and $x_1 = 0.1$ and

$$\left(\frac{dy}{dx}\right)_1 = -(0.8)(0.1) - 2 = -2.08$$

Hence $y_2 \approx h\left[\left(\frac{dy}{dx}\right)_1\right] + y_1 \approx 0.1(-2.08) + 0.8$

That is: $y_2 \approx 0.592$

Step 3
Finally, take $y_2 = 0.592$, $x_2 = 0.2$

and $\left(\frac{dy}{dx}\right)_2 = -(0.592)(0.2) - 2 = -2.1184$

So: $y_3 \approx h\left[\left(\frac{dy}{dx}\right)_2\right] + y_2$

$$\approx 0.1(-2.1184) + 0.592 = 0.3802 \quad \text{(4 decimal places)}$$

Did you spot that a recurrence relation is being used here? This relation is

$$y_{n+1} = h(-x_n y_n - 2) + y_n$$

and, in your own calculations, you should show your workings in a table like this:

n	x_n	y_n	$y_n - 0.1(x_n y_n + 2)$
0	0	1	0.8
1	0.1	0.8	0.592
2	0.2	0.592	0.3802
3	0.3	0.3802	

You should understand that there is no guarantee that the values $y_1, y_2, y_3, \ldots$ which you obtain are reliable. Further advanced study beyond your present syllabus would be required before you could decide to either accept or reject any estimate. In examination questions, you will be asked to apply a formula such as $\left(\dfrac{dy}{dx}\right)_0 \approx \dfrac{y_1 - y_0}{h}$ once, or successively, to obtain a particular result from given data.

Euler's method is often adapted in the following way. From Taylor's expansion for values of x near to x_0 you have

$$f(x) \approx f(x_0) + (x - x_0)\,f'(x_0)$$

where powers of $(x - x_0)$ higher than the first are neglected. Suppose h is small, then at $x = x_0 - h$

$$f(x_0 - h) \approx f(x_0) + (x_0 - h - x_0)\,f'(x_0)$$

That is: $\qquad\qquad f(x_0 - h) \approx f(x_0) - h f'(x_0)$ $\qquad\qquad$ (1)

Also at $x = x_0 + h$, you have similarly

$$f(x_0 + h) \approx f(x_0) + h f'(x_0) \qquad\qquad (2)$$

Subtracting equation (1) from equation (2) gives

$$f(x_0 + h) - f(x_0 - h) \approx 2h f'(x_0)$$

Write $f(x_0 - h) = y_{-1}$, $f'(x_0) = \left(\dfrac{dy}{dx}\right)_0$ and $f(x_0 + h) = y_1$ and you have the formula

$$y_1 - y_{-1} \approx 2h\left(\frac{dy}{dx}\right)_0$$

which gives on rearranging

∎

$$\left(\frac{dy}{dx}\right)_0 \approx \frac{y_1 - y_{-1}}{2h}$$

The formula can be illustrated geometrically like this:

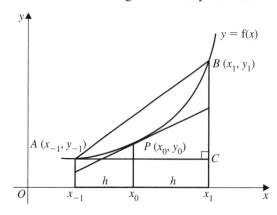

The point P has coordinates (x_0, y_0) on the curve with equation $y = f(x)$. The point A on the curve has coordinates (x_{-1}, y_{-1}) where $x_0 - x_{-1} = h$. The point B on the curve has coordinates (x_1, y_1) where $x_1 - x_0 = h$. The gradient of the tangent to the curve at P is $f'(x_0)$ which you can write as $\left(\dfrac{dy}{dx}\right)_0$. Now, again, for small h the gradient of the chord AB is approximately equal to the gradient of the tangent to the curve at P. But $BC = y_1 - y_{-1}$ and hence the gradient of the chord $AB = \dfrac{y_1 - y_{-1}}{2h}$.

Hence you have the formula

$$\left(\frac{dy}{dx}\right)_0 \approx \frac{y_1 - y_{-1}}{2h}$$

Since also $\left(\dfrac{dy}{dx}\right)_0 = g(x_0, y_0)$ from the differential equation, you have

$$\frac{y_1 - y_{-1}}{2h} \approx g(x_0, y_0)$$

That is:
$$y_1 \approx y_{-1} + 2hg(x_0, y_0)$$

or, more generally, $\quad y_{n+1} \approx y_{n-1} + 2hg(x_n, y_n)$

When you are working exercises, you are usually given $g(x, y)$ and the value, y_0, of y at $x = x_0$. The adapted Euler formula, however, requires a 'double step' and the value of y at x_{-1} is not known. Instead you need to use Euler's formula $y_1 \approx y_0 + h\left(\dfrac{dy}{dx}\right)_0$ (or some other method) to get an estimate of y_1. This in effect then means that the formula $y_1 \approx y_{-1} + 2hg(x_0, y_0)$ is initially bypassed because you already have a value for y_1.

Now, however, you can use $y_{n+1} \approx y_{n-1} + hg(x_n, y_n)$ to obtain:

$$y_2 \approx y_0 + 2hg(x_1, y_1)$$
$$y_3 \approx y_1 + 2hg(x_2, y_2)$$

and so on, to the number of steps required in a particular question. This adaption of Euler's method usually gives greater accuracy in the estimates for y_2, y_3, etc. You can also estimate y_1 by using a Taylor series at the start. The differential equation used in example 1 is now solved by using this adaptation of Euler's method in example 2.

Example 2

$$\frac{dy}{dx} = -2 - xy \quad \text{and} \quad y = 1 \quad \text{at} \quad x = 0$$

Use the formula $y_1 \approx y_0 + h\left(\dfrac{dy}{dx}\right)_0$, with $h = 0.1$, to find an estimate of y at $x = 0.1$.

Using a step length of 0.1, use the formula

$$\left(\frac{dy}{dx}\right)_1 \approx \frac{y_2 - y_0}{2h}$$

as many times as required to find an estimate of y at $x = 0.3$, giving 4 decimal places in your answer.

Step 1
From data in the question, $x_0 = 0$, $y_0 = 1$, $h = 0.1$ and

$$\left(\frac{dy}{dx}\right)_0 = -2 - x_0 y_0 = -2$$

Then
$$y_1 \approx y_0 + h\left(\frac{dy}{dx}\right)_0 = 1 + (0.1)(-2) = 0.8$$

Step 2
You now have $x_1 = 0.1$, $y_1 = 0.8$ which gives

$$\left(\frac{dy}{dx}\right)_1 = -2 - x_1 y_1 = -2 - 0.08 = -2.08$$

Using $y_2 - y_0 \approx 2h\left(\frac{dy}{dx}\right)_1$, you have

$$y_2 - 1 \approx 0.2(-2.08)$$
$$y_2 \approx 1 - 0.416 = 0.584$$

Step 3
You now have $x_2 = 0.2$, $y_2 = 0.584$ which gives

$$\left(\frac{dy}{dx}\right)_2 = -2 - x_2 y_2 = -2 - 0.1168 = -2.1168$$

Using $y_3 - y_1 \approx 2h\left(\frac{dy}{dx}\right)_2$, you have

$$y_3 - 0.8 \approx 2(0.1)(-2.1168)$$
$$y_3 \approx 0.8 - 0.42336 = 0.376\,64$$

That is, at $x = 0.3$, the estimated value of y by this method is 0.3766 (4 decimal places).

The table shows a summary of the calculation. Remember you need to obtain an estimate of y_1 before using the iterative formula

$$y_{n+1} \approx y_{n-1} + 0.2(-2 - x_n y_n)$$

which comes from combining with $h = 0.1$ the two formulae,

$$\frac{y_{n+1} - y_{n-1}}{2h} \approx \left(\frac{dy}{dx}\right)_n \quad \text{and} \quad \left(\frac{dy}{dx}\right)_n = -2 - x_n y_n.$$

n	x_n	y_n	$y_{n-1} + 0.2(-2 - x_n y_n)$
0	0	1	
1	0.1	0.8 (estimate)	
2	0.2	0.584 ⟵	⟶ 0.584
3	0.3	0.3766(4)	0.376 64

Step-by-step methods can be found also for solving second order differential equations. Here you need a formula for $\frac{d^2 y}{dx^2}$ at $x = x_0$.

The first three terms of the Taylor expansion of $f(x)$ in ascending powers of $(x - x_0)$ are used like this:

$$f(x) \approx f(x_0) + (x - x_0) f'(x_0) + \frac{(x - x_0)^2}{2!} f''(x_0)$$

Suppose that h is small, then at $x = x_0 - h$ you have

$$f(x_0 - h) \approx f(x_0) - hf'(x_0) + \frac{h^2}{2!} f''(x_0) \tag{1}$$

Also at $x = x_0 + h$ you have

$$f(x_0 + h) \approx f(x_0) + hf'(x_0) + \frac{h^2}{2!} f''(x_0) \tag{2}$$

Adding equations (1) and (2):

$$f(x_0 - h) + f(x_0 + h) \approx 2f(x_0) + h^2 f''(x_0)$$

Writing $\quad f(x_0 - h) = y_{-1}, \quad f(x_0) = y_0, \quad f(x_0 + h) = y_1 \quad$ and $f''(x_0) = \left(\frac{d^2 y}{dx^2}\right)_0$ you have

$$y_1 + y_{-1} \approx 2y_0 + h^2 \left(\frac{d^2 y}{dx^2}\right)_0$$

■ **That is:** $\quad \left(\frac{d^2 y}{dx^2}\right)_0 \approx \frac{y_1 - 2y_0 + y_{-1}}{h^2}$

Note: If a second order differential equation contains a term in $\frac{dy}{dx}$, you may need to use the formula $\left(\frac{dy}{dx}\right)_0 \approx \frac{y_1 - y_{-1}}{2h}$ in addition to the formula for $\left(\frac{d^2 y}{dx^2}\right)_0$ shown above. Here is an example to show you how this works.

Example 3

$$\frac{d^2y}{dx^2} = x + y^2, \frac{dy}{dx} = y = 1 \quad \text{at} \quad x = 0$$

Using a step length $h = 0.1$ and the formulae

$$\left(\frac{dy}{dx}\right)_0 \approx \frac{y_1 - y_{-1}}{2h}, \left(\frac{d^2y}{dx^2}\right)_0 \approx \frac{y_1 - 2y_0 + y_{-1}}{h^2}$$

find estimates for the value of y at $x = 0.1, 0.2$ and 0.3.

Step 1
Take $x_0 = 0$, then $y_0 = 1$, y_{-1} is the value of y at $x = -0.1$ and y_1 is the value of y at $x = 0.1$.

$$\left(\frac{dy}{dx}\right)_0 = 1, \left(\frac{d^2y}{dx^2}\right)_0 = 0 + 1^2 = 1. \text{ Also } h = 0.1.$$

Substituting into $\left(\frac{dy}{dx}\right)_0 \approx \frac{y_1 - y_{-1}}{2h}$ you have

$$1 \approx \frac{y_1 - y_{-1}}{0.2} \Rightarrow y_1 - y_{-1} \approx 0.2 \qquad (1)$$

Substituting into $\left(\frac{d^2y}{dx^2}\right)_0 \approx \frac{y_1 - 2y_0 + y_{-1}}{h^2}$ you have

$$1 \approx \frac{y_1 - 2 + y_{-1}}{0.01} \Rightarrow y_1 + y_{-1} \approx 2.01 \qquad (2)$$

Add equations (1) and (2) and you have
$$2y_1 \approx 2.21 \Rightarrow y_1 \approx 1.105$$

Step 2
You have $x_1 = 0.1$, $y_1 = 1.105$ and $y_0 = 1$.
From the differential equation $\left(\frac{d^2y}{dx^2}\right)_1 = x_1 + y_1^2$:

$$\left(\frac{d^2y}{dx^2}\right)_1 = 0.1 + 1.105^2$$

Using the formula $\left(\frac{d^2y}{dx^2}\right)_1 \approx \frac{y_2 - 2y_1 + y_0}{h^2}$ you have

$$0.1 + 1.105^2 \approx \frac{y_2 - 2(1.105) + 1}{0.01}$$

So: $\quad\quad y_2 \approx 1.223\,210\,3 \rightarrow 1.2232$ (4 decimal places)

Step 3

The process shown in step 2 is now repeated with $x_2 = 0.2$, $y_2 = 1.223\,2103$ and $y_1 = 1.105$.

$$\left(\frac{d^2y}{dx^2}\right)_2 = x_2 + y_2^2 \approx 0.2 + 1.223\,210\,3^2$$

Using the formula $\left(\dfrac{d^2y}{dx^2}\right)_2 \approx \dfrac{y_3 - 2y_2 + y_1}{h^2}$ you have

$$0.2 + 1.223\,210\,3^2 \approx \frac{y_3 - 2(1.223\,210\,3) + 1.105}{0.01}$$

So: $\qquad\qquad\qquad y_3 \approx 1.3584$ (4 decimal places)

Note: The first five terms in the Taylor series solution in ascending powers of x of the differential equation $\dfrac{d^2y}{dx^2} = x + y^2$ for which $y = \dfrac{dy}{dx} = 1$ at $x = 0$ are

$$1 + x + \tfrac{1}{2}x^2 + \tfrac{1}{2}x^3 + \tfrac{1}{6}x^4$$

The values of y_1, y_2, y_3 obtained in the step-by-step solution compare well with those from the series solution and you should investigate this in detail for yourself.

The table below shows a summary of the calculation in Example 3. By combining $\left(\dfrac{d^2y}{dx^2}\right)_n \approx \dfrac{y_{n+1} - 2y_n + y_{n-1}}{h^2}$ with the differential equation at (x_n, y_n) which can be written as $\left(\dfrac{d^2y}{dx^2}\right)_n = x_n + y_n^2$ you have

$$\frac{y_{n+1} - 2y_n + y_{n-1}}{h^2} \approx x_n + y_n^2, \text{ where } h = 0.1$$

That is, $y_{n+1} \approx 0.01\left(x_n + y_n^2\right) + 2y_n - y_{n-1}$ is the iterative relation underpinning the calculation.

n	x_n	y_n	$0.01\left(x_n + y_n^2\right) + 2y_n - y_{n-1}$
0	0	1	
1	0.1	1.105	1.2232
2	0.2	1.2232	1.3584
3	0.3	1.3584	

Exercise 5A

In questions 1–6, use the formula $\left(\dfrac{dy}{dx}\right)_0 \approx \dfrac{y_1 - y_0}{h}$.

1 $\dfrac{dy}{dx} = 3x^2y + 1, \quad y = 1$ at $x = 0$

Using a step length of 0.1, find y at $x = 0.2$.

2 $\dfrac{dy}{dx} = \ln(x + y), \quad y = 2$ at $x = 0$

Using a step length of 0.1, find y at $x = 0.2$.

3 $\dfrac{dy}{dx} = \sin x^2, \quad y = 2$ at $x = 1$

Using a step length of 0.2, find y at $x = 1.4$.

4 $\dfrac{dy}{dx} = x^2 - y^2, \quad y = 0$ at $x = 0$

Using a step length of 0.2, find y at $x = 0.8$.

5 $\dfrac{dy}{dx} = \dfrac{4x^2 + y^2}{x + y}, \quad y = 4$ at $x = 1$

Using a step length of 0.2, find y at $x = 1.4$.

6 $\dfrac{dy}{dx} = \dfrac{e^{x+y}}{x + y}, \quad y = 1$ at $x = 0.8$

Using a step length of 0.1, find y at $x = 1$.

In questions 7–12, first use the formula $y_1 \approx y_0 + h\left(\dfrac{dy}{dx}\right)_0$ to find an estimate of y_1 and then use the formula $\left(\dfrac{dy}{dx}\right)_1 \approx \dfrac{y_2 - y_0}{2h}$ as many times as required, giving your final answer to 4 decimal places.

7 $\dfrac{dy}{dx} = 1 + \ln x, \quad y = 0$ at $x = 1$

Using a step length of 0.1, find y at $x = 1.3$.

8 $\dfrac{dy}{dx} = y + 2e^x, \quad y = 0$ at $x = 0$

Using a step length of 0.2, find y at $x = 0.6$.

9 $\dfrac{dy}{dx} = xy, \quad y = 1$ at $x = 0$

Using a step length of 0.1, find y at $x = 0.3$.

10 $\dfrac{dy}{dx} = \dfrac{3x^2 - y^2}{2xy}, \quad y = 2$ at $x = 1$

Using a step length of 0.1, find y at $x = 1.3$.

11 $\dfrac{dy}{dx} = 2 + x + \sin y, \quad y = 0$ at $x = 0$

Using a step length of 0.05, find y at $x = 0.2$.

12 $\dfrac{dy}{dx} = x^2 - y^2, \quad y = 1$ at $x = 2$

Using a step length of 0.1, find y at $x = 2.3$.

In questions 13–20, use the formula $\left(\dfrac{d^2y}{dx^2}\right)_0 \approx \dfrac{y_1 - 2y_0 + y_{-1}}{h^2}$ and,

when necessary, the formula $\left(\dfrac{dy}{dx}\right)_0 \approx \dfrac{y_1 - y_{-1}}{2h}$ to estimate solutions

to the following differential equations, giving final answers to 4 decimal places.

13 $\dfrac{d^2y}{dx^2} = y - x - 1, \; \dfrac{dy}{dx} = 2$ and $y = 0$ at $x = 0$

Using a step length of 0.1, find y at $x = 0.3$.

14 $\dfrac{d^2y}{dx^2} = \dfrac{4x - 4y^2 - 4x^2 - 1}{4y^3}$ for which $y = 0.3$ at $x = 0.1$

and $y = 0.4$ at $x = 0.2$

Using a step length of 0.1, estimate the value of y at
(a) $x = 0.3$ (b) $x = 0.4$.

15 $\dfrac{d^2y}{dx^2} = 1 + \sinh x \sinh y, \; \dfrac{dy}{dx} = y = 1$ at $x = 0$

Use a step length of 0.1 to estimate the value of y at
(a) $x = 0.2$ (b) $x = 0.3$.

16 $\dfrac{d^2y}{dx^2} + 10\dfrac{dy}{dx} - y^2 = 0, \; y = 1$ at $x = 0$ and $y = 2$ at $x = 0.1$

Use a step length of 0.1 to estimate the value of y at $x = 0.3$.

17 $\dfrac{d^2y}{dx^2} = 1 + \tfrac{1}{2}y\dfrac{dy}{dx}, \; y = 2$ at $x = 0$ and $y = 2.1$ at $x = 0.1$

Use a step length of 0.1 to estimate the value of y at (a)
$x = 0.2$ (b) $x = 0.3$, giving answers to 3 decimal places.

18 $\dfrac{d^2y}{dx^2} = 1 + x \cos y + \sin y \cos y, \; \dfrac{dy}{dx} = 1$ and $y = 0$ at $x = 1$

Using a step length of 0.05, estimate the value of y at
(a) $x = 1.05$ (b) $x = 1.1$.

19 $\dfrac{d^2y}{dx^2} = 2x + 2x^2y + 2y^3$, $\dfrac{dy}{dx} = 8$ and $y = 2$ at $x = 2$

Using a step length of 0.2, estimate the value of y at

(a) $x = 2.2$ (b) $x = 2.4$.

20 $\dfrac{d^2y}{dx^2} - \dfrac{dy}{dx} - 3x^2 = 0$, $\dfrac{dy}{dx} = 1$ and $y = 2$ at $x = 0$

Using a step length of 0.1, estimate the value of y at $x = 0.2$.
Check the accuracy of your answer by finding the series
solution for y.

SUMMARY OF KEY POINTS

1 In step-by-step methods, where the step length is h, learn
the approximations and how to use them:

$$\left(\frac{dy}{dx}\right)_0 \approx \frac{y_1 - y_0}{h}$$

$$\left(\frac{dy}{dx}\right)_0 \approx \frac{y_1 - y_{-1}}{2h}$$

$$\left(\frac{d^2y}{dx^2}\right)_0 \approx \frac{y_1 - 2y_0 + y_{-1}}{h^2}$$

Proof

In any attempt at writing a proof in mathematics you are trying to establish whether a result or a proposition is true or not true. In this chapter you are shown how to establish results by using a proof based on **mathematical induction**.

6.1 Proof by induction

Many theorems and formulae relating to positive whole numbers can be proved by a process called **mathematical induction**. The method can be described like this.

A theorem thought to be true for all values of the positive integer n can be proved by showing that

(i)   if it is true for $n = k$, then it is also true for $n = k + 1$
(ii) it is true for some small value of n such as $n = 1$ (or perhaps $n = 2$ or 3).

If you can prove both (i) and (ii), then you have shown that the theorem is true at the start ($n = 1$, say) and it is therefore true for $n = 1 + 1$, and then $n = 2 + 1$, and then $n = 3 + 1$, and so on for all integral values of n following after the valid starting value (usually $n = 1$, but not always). This way of proving the validity of a theorem or formula is the method of mathematical induction.

An essential requirement when trying to prove a proposition by induction is that you either know the final result or can make an inspired guess and then confirm your guess by using a proof employing induction. Consider the following identities for hyperbolic functions, which you can easily prove by direct deduction:

$$(\cosh\theta + \sinh\theta)^2 \equiv \cosh 2\theta + \sinh 2\theta$$
$$(\cosh\theta + \sinh\theta)^3 \equiv \cosh 3\theta + \sinh 3\theta$$

and it is clear also that

$$(\cosh\theta + \sinh\theta)^1 \equiv \cosh\theta + \sinh\theta$$

These results suggest that it may be worth investigating whether $(\cosh\theta + \sinh\theta)^n$ is identically the same as $\cosh n\theta + \sinh n\theta$ for all positive integral values of n.

Example 1

Use the method of mathematical induction to prove that

$$(\cosh\theta + \sinh\theta)^n \equiv \cosh n\theta + \sinh n\theta$$

where n is a positive integer.

Assume that the identity is true for $n = k$.

That is: $\qquad (\cosh\theta + \sinh\theta)^k \equiv \cosh k\theta + \sinh k\theta$

Now $\quad (\cosh\theta + \sinh\theta)^{k+1} \equiv (\cosh\theta + \sinh\theta)(\cosh\theta + \sinh\theta)^k$
$$= (\cosh\theta + \sinh\theta)(\cosh k\theta + \sinh k\theta)$$

by using the assumption made.

Hence:

$$(\cosh\theta + \sinh\theta)^{k+1} \equiv \cosh\theta\cosh k\theta + \sinh\theta\sinh k\theta + \sinh\theta\cosh k\theta + \cosh\theta\sinh k\theta$$

But you know that $\quad \cosh(A + B) \equiv \cosh A \cosh B + \sinh A \sinh B$
and $\qquad\qquad\qquad \sinh(A + B) \equiv \sinh A \cosh B + \cosh A \sinh B$

(Book P5, page 12).

So: $\qquad (\cosh\theta + \sinh\theta)^{k+1} \equiv \cosh(k + 1)\theta + \sinh(k + 1)\theta$

So if the identity is true for $n = k$, then it is true for $n = k + 1$. You also know by inspection that the identity is true when $n = 1$.

Hence, by mathematical induction, the identity

$$(\cosh\theta + \sinh\theta)^n \equiv \cosh n\theta + \sinh n\theta$$

is true for all positive, integral n.

Example 2

Use the method of mathematical induction to prove that

$$\sum_{r=1}^{n} r^3 = \tfrac{1}{4}n^2(n + 1)^2$$

where n is a positive integer.

Assume that the summation is true for $n = k$.

That is: $\qquad 1^3 + 2^3 + 3^3 + \ldots + k^3 = \tfrac{1}{4}k^2(k + 1)^2$

The next term in the series is $(k+1)^3$ and on adding this term to both sides you have

$$1^3 + 2^3 + 3^3 + \ldots + k^3 + (k+1)^3 = \tfrac{1}{4}k^2(k+1)^2 + (k+1)^3$$

That is:
$$\sum_{r=1}^{k+1} r^3 = \tfrac{1}{4}(k+1)^2\left[k^2 + 4(k+1)\right]$$

$$= \tfrac{1}{4}(k+1)^2(k^2 + 4k + 4)$$

$$= \tfrac{1}{4}(k+1)^2(k+2)^2$$

$$= \tfrac{1}{4}(k+1)^2\left[(k+1)+1\right]^2$$

But this result is the same as that which would be obtained by replacing k by $k+1$ in the formula assumed as $\sum_{r=1}^{k} r^3 = \tfrac{1}{4}k^2(k+1)^2.$

So if the summation is true for $n=k$ it is also true for $n=k+1$.

For $n=1$, the left-hand side is $1^3 = 1$ and the right-hand side is $\tfrac{1}{4}(1^2)(1+1)^2 = 1$.

So the summation is true for $n=1$.

Therefore it is true for $n=1+1$, $n=2+1$, $n=3+1$, and so on.

That is, by mathematical induction:

$$\sum_{r=1}^{n} r^3 = \tfrac{1}{4}n^2(n+1)^2$$

for all positive, integral values of n.

Example 3
Use the method of mathematical induction to prove that the expression $3^{2n} + 7$ is divisible by 8 for all positive, integral values of n.

Let $f(k) = 3^{2k} + 7$, then $f(k+1) = 3^{2(k+1)} + 7$ and you have

$$f(k+1) - f(k) = 3^{2k+2} - 3^{2k}$$
$$= 9(3^{2k}) - 3^{2k}$$

because $\qquad 3^{2k+2} = (3^{2k})(3^2) = 9(3^{2k})$

Hence $\qquad f(k+1) = f(k) + 8(3^{2k})$

This implies that if $f(n)$ is divisible by 8 for some integral value of $n=k$, then $f(n)$ is also divisible by 8 for $n=k+1$.

For $n=1$, $f(n) = f(1) = 3^2 + 7 = 16$ which is divisible by 8.

So $f(1 + 1) = f(2)$ is divisible by 8.

and $f(1 + 2) = f(3)$ is divisible by 8 and so on ...

By mathematical induction $3^{2n} + 7$ is divisible by 8 for all positive integral n.

Example 4

Given that n is an integer which is greater than 3, show that $n! > 2^n$.

For $n = 4$,
$$n! = 4! = 24$$
$$2^n = 2^4 = 16$$

So for $n = 4$, it is true that $n! > 2^n$.

Assume that for some integer k, greater than 4, you have $k! > 2^k$.

Multiply by $k + 1$ to obtain

$$k!(k + 1) > (k + 1)2^k$$

That is:
$$(k + 1)! > (k + 1)2^k$$

Since $k > 4$, you have $k + 1 > 5$ and so $k + 1$ is certainly > 2.

So:
$$(k + 1)! > (2)(2^k) = 2^{k+1}$$

If the inequality $n! > 2^n$ is true for $n = k$, it is also true for $n = k + 1$, where $k \geqslant 4$. But the inequality is true for $n = 4$, so it is true for $n = 5, 6, \ldots$ Therefore by induction $n! > 2^n$ is true for all integral $n \geqslant 4$.

Example 5

Prove, by induction, that if $u_{n+2} = 5u_{n+1} - 6u_n$ with $u_1 = 1$ and $u_2 = 5$, then $u_n = 3^n - 2^n$.

Assume that it is true for $n = k$ and $n = k + 1$, that is:

$$u_k = 3^k - 2^k$$

and
$$u_{k+1} = 3^{k+1} - 2^{k+1}$$

Then:
$$
\begin{aligned}
u_{k+2} &= 5(3^{k+1} - 2^{k+1}) - 6(3^k - 2^k) \\
&= 5(3^{k+1}) - 6(3^k) - 5(2^{k+1}) + 6(2^k) \\
&= 5(3^{k+1}) - 2(3^{k+1}) - 5(2^{k+1}) + 3(2^{k+1}) \\
&= 3^{k+1}(5 - 2) - 2^{k+1}(5 - 3) \\
&= 3^{k+1} \cdot 3 - 2^{k+1} \cdot 2
\end{aligned}
$$

So
$$u_{k+2} = 3^{k+2} - 2^{k+2}$$

Hence if the result is true for $n = k$ and $n = k + 1$ then it is also true for $n = k + 2$.

Now $u_1 = 3^1 - 2^1 = 1$, as given, so it is true for $n = 1$.

and $u_2 = 3^2 - 2^2 = 9 - 4 = 5$, as given, so it is true for $n = 2$.

Thus: $$u_3 = 5u_2 - 6u_1$$
$$= (5 \times 5) - (6 \times 1) = 25 - 6 = 19$$

and $$19 = 3^3 - 2^3$$

So it is true for $n = 1$ and $n = 2$.

Thus it is true for $n = 3$.

But if it is true for $n = 2$ and $n = 3$ then it is true for $n = 4$, and so on.

Consequently, $u_n = 3^n - 2^n$ for all positive integers.

Exercise 6A

In questions 1–15, use the method of mathematical induction to prove the result given.

1 $\displaystyle\sum_{r=1}^{n} r = \frac{1}{2}n(n+1)$

2 $\displaystyle\sum_{r=1}^{n} r^2 = \frac{1}{6}n(n+1)(2n+1)$

3 $\displaystyle\sum_{r=1}^{n} r(r+1) = \frac{1}{3}n(n+1)(n+2)$

4 $\displaystyle\sum_{r=1}^{n} r(r!) = (n+1)! - 1$

5 $\displaystyle\sum_{r=1}^{n} \frac{1}{(r+1)(r+2)} = \frac{n}{2(n+2)}$

6 $1 + 2x + 3x^2 + \ldots + nx^{n-1} = \dfrac{1-x^n}{(1-x)^2} - \dfrac{nx^n}{1-x}$

7 $\displaystyle\sum_{r=1}^{n} r(3r-1) = n^2(n+1)$

8 $\displaystyle\sum_{r=1}^{n} \sin^2(2r-1)\theta = \frac{1}{2}n - \frac{\sin 4n\theta}{4\sin 2\theta}$

9 $\displaystyle\sum_{r=1}^{n} \frac{3^r(r+1)}{(r+4)!} = \frac{1}{8} - \frac{3^{n+1}}{(n+4)!}$

10 $\displaystyle\sum_{r=1}^{n} r(r+1)^2 = \frac{1}{12}n(n+1)(n+2)(3n+5)$

11 $\displaystyle\sum_{r=1}^{n} \frac{2r-1}{2^{r-1}} = 6 - \frac{2n+3}{2^{n-1}}$

12 $\displaystyle\sum_{r=1}^{n} \sin(2r-1)\theta = \frac{\sin^2 n\theta}{\sin\theta}$

13 $\displaystyle\sum_{r=1}^{n} \cos(2r-1)\theta = \frac{\sin n\theta \cos n\theta}{\sin\theta}$

14 $\displaystyle\sum_{r=1}^{n} \operatorname{cosec}(2^r\theta) = \cot\theta - \cot(2^n\theta)$

15 $\displaystyle\sum_{r=1}^{n} \tan r\theta \tan(r+1)\theta = \tan(n+1)\theta \cot\theta - n - 1$

16 Given that n is a positive integer, prove that

$$(\cos\theta + \mathrm{i}\sin\theta)^n = \cos n\theta + \mathrm{i}\sin n\theta$$

17 Given that $^nC_r = \dfrac{n!}{r!\,(n-r)!}$, show that

$$^{n+1}C_r = {}^nC_r + {}^nC_{r-1}$$

Use the method of mathematical induction to prove that

$$(1+x)^n = {}^nC_0 + {}^nC_1 x + {}^nC_2 x^2 + \ldots + {}^nC_n x^n$$

where n is a positive integer.

18 Use the method of induction to show that

$$\sum_{r=1}^{2n} r^3 = n^2(2n+1)^2$$

19 Given that n is a positive integer, prove that $n(n+1)(2n+1)$ is divisible by 6.

20 Given that n is a positive integer, prove that $3^{4n+2} + 2^{6n+3}$ is divisible by 17.

21 (a) If n is an odd positive integer, prove that $2^n + 1$ is divisible by 3.
(b) If n is an even positive integer, prove that $2^n - 1$ is divisible by 3.

22 Given that n is an even positive integer, prove that $(2^{2n} - 1)$ is divisible by 5.

23 Given that n is an odd positive integer, prove that $(5^{2n} + 1)$ is divisible by 13.

24 Given that $A_n = 2^{n+2} + 3^{2n+1}$, show that

$$A_{n+1} - 2A_n = 7\left(3^{2n+1}\right)$$

Hence use the method of mathematical induction to prove that A_n is divisible by 7, where n is any positive integer.

25 Given that m is an odd positive integer, prove that $(m^2 + 3)(m^2 + 15)$ is divisible by 32 for all such values of m.

26 Given that n is a positive integer, prove that $3^{2n} + 11$ is divisible by 4.

27 Given that n is a positive integer, prove that $(3n + 1)7^n - 1$ is divisible by 9.

28 Given that $0 < x < \frac{\pi}{2}$, and n is a positive integer, prove that $(1 - \sin x)^n < 1$.

29 Given that n is a positive integer, use the method of mathematical induction to prove that

$$\sum_{r=1}^{n} r^2 \geqslant n\left(\frac{n+1}{2}\right)^2$$

30 Given that n is a positive integer, prove by induction that

$$1 + 2 + 3 + \ldots + n > \tfrac{1}{2}n^2$$

31 Given that n is a positive integer, prove that

$$\tfrac{1}{2} + \tfrac{3}{4} + \tfrac{5}{6} + \ldots + \frac{2n-1}{2n} < n - \tfrac{1}{2}, \text{ for } n \geqslant 2$$

32 Given that n is a positive integer, prove that

$$\tfrac{1}{2} + \tfrac{3}{4} + \tfrac{5}{6} + \ldots + \frac{2n-1}{2n} > \tfrac{1}{2}n, \text{ for } n \geqslant 2$$

33 Given that n is a positive integer, show by the method of induction that

$$\frac{n}{2} < \tfrac{1}{2} + \tfrac{2}{3} + \tfrac{3}{4} + \ldots + \frac{n}{n+1} < n, \text{ for } n \geqslant 2$$

34 Given that $u_{n+1} = 2u_n + 1$, $u_1 = 3$, prove by induction that

$$u_n = 2^{n+1} - 1$$

35 Given that $u_{n+1} = 3u_n + 4$, $u_1 = 1$, prove by induction that

$$u_n = 3^n - 2$$

36 Given that $u_{n+2} = 7u_{n+1} - 10u_n$, $u_1 = -1$, $u_2 = 13$, prove by induction that

$$u_n = 5^n - 3(2^n)$$

37 Given that $u_{n+2} - 4u_{n+1} + 4u_n - n = 0$, $u_1 = 1$, $u_2 = 2$, prove by induction that

$$u_n = n + 2 + (n-3)2^{n-1}$$

38 Given that $u_{n+1} - 4u_n + 5u_{n-1} - 2u_{n-2} = 0$, $u_1 = 3$, $u_2 = 1$, $u_3 = 0$, prove that

$$u_n = 2^{n-1} - 3n + 5$$

39 Given that n is a positive integer where $n \geqslant 2$, prove by the method of mathematical induction that

(a) $\displaystyle\sum_{r=1}^{n-1} r^3 < \frac{n^4}{4}$

(b) $\displaystyle\sum_{r=1}^{n} r^3 > \frac{n^4}{4}$

SUMMARY OF KEY POINTS

A proof by mathematical induction consists in showing that if a theorem is true for some special integral value of n, say $n = k$, then it is true for $n = k + 1$. Also you need to show that the theorem is true for some trivial value of n such as $n = 1$ (or $n = 2$, etc.). Then if it is true for $n = k + 1$, when it is true for $n = k$, and if it is true for $n = 1$, then it is true for $n = 1 + 1 = 2$, $n = 1 + 2 = 3$ and so on for all positive integral n.

Review exercise

1 Solve the equation

$$z^3 + 1 = 0$$

giving your solutions in the form $a + ib$, where a and b are real, and showing them on an Argand diagram. [E]

2 Provided that x is so small that terms in x^3 and higher powers of x may be neglected, show that

$$4\sqrt{(1+x)} = 3 + 2x + \cos x$$ [E]

3 The point $P(x, y, z)$ is mapped on the point $Q(X, Y, Z)$ by the relation

$$\begin{pmatrix} X \\ Y \\ Z \end{pmatrix} = \mathbf{M} \begin{pmatrix} x \\ y \\ z \end{pmatrix}, \text{ where } \mathbf{M} = \begin{pmatrix} -1 & 2 & 0 \\ 2 & 0 & -2 \\ 0 & -2 & 1 \end{pmatrix}$$

Show that Q lies on the plane with equation $2X + Y + 2Z = 0$.
Show also that $\mathbf{M}^3 = \lambda \mathbf{M}$ and give the value of λ. [E]

4 The position vectors $\mathbf{a}$ and $\mathbf{b}$ of points A and B are $\mathbf{i} - \mathbf{k}$ and $3\mathbf{i} + \lambda\mathbf{j} - \mathbf{k}$ respectively, where λ is a scalar.
(a) Find $\mathbf{a} \times \mathbf{b}$ in terms of λ.
Given that the area of $\triangle OAB$ is $\sqrt{3}$,
(b) find the possible values of λ. [E]

5 By using the series expansions of e^x and $\cos x$, or otherwise, find the expansion of $e^x \cos 3x$ in ascending powers of x up to and including the term in x^3. [E]

6 Given that $\left| \dfrac{z-1}{z+1} \right| = 2$, find the cartesian equation of the locus of z and represent the locus by a sketch in the Argand diagram.
Shade the region for which the inequalities $\left| \dfrac{z-1}{z+1} \right| > 2$ and $0 < \arg z < \frac{3\pi}{4}$ are both satisfied. [E]

7 With respect to a fixed origin O, the point A has position vector $2\mathbf{i} - \mathbf{j} + 3\mathbf{k}$. The line l passes through A and is parallel to the vector $-\mathbf{i} + 2\mathbf{j} - 2\mathbf{k}$.

(a) Giving your answer to the nearest degree, calculate the acute angle between OA and l.

The point B with position vector $3\mathbf{i} + u\mathbf{j} + v\mathbf{k}$ lies on l.

(b) Calculate the values of u and v. [E]

8 Given that $|2x| < 1$, find the first two non-zero terms in the expansion of

$$\ln[(1 + x)^2(1 - 2x)]$$

in a series of ascending powers of x. [E]

9
$$\mathbf{A} = \begin{pmatrix} 1 & 0 & 2 \\ t & 3 & 1 \\ -2 & -1 & 1 \end{pmatrix}$$

Given that $\mathbf{A}$ is singular, find the value of t. [E]

10 Find the modulus and argument of the complex number

$$\frac{5 + i}{3 - 2i}$$

Hence solve the equation

$$z^4 = \frac{5 + i}{3 - 2i}$$

giving your answers in the form $re^{i\theta}$, where $r > 0$ and $-\pi < \theta \leqslant \pi$. [E]

11 Find the eigenvalues of the matrix $\mathbf{A}$ where

$$\mathbf{A} = \begin{pmatrix} 4 & -2 & 0 \\ -2 & 0 & -2 \\ 0 & -2 & 1 \end{pmatrix}$$

Given that the matrix $\mathbf{P}$ is such that $\mathbf{P}^{-1}\mathbf{AP}$ is a diagonal matrix $\mathbf{B}$, write down a possible form for $\mathbf{B}$. [E]

12 If $z = \cos\theta + i\sin\theta$, show that

$$z - \frac{1}{z} = 2i\sin\theta, \quad z^n - \frac{1}{z^n} = 2i\sin n\theta$$

Hence, or otherwise, show that

$$16\sin^5\theta = \sin 5\theta - 5\sin 3\theta + 10\sin\theta \qquad [E]$$

13 (a) Show that the first four terms of the expansion of $e^{\tan x}$ in a series of ascending powers of x are

$$1 + x + \tfrac{1}{2}x^2 + \tfrac{1}{2}x^3$$

(b) Deduce the first four terms of the expansion of $e^{-\tan x}$ in a series of ascending powers of x.   [E]

14 Given that x is so small that terms in x^3 and higher powers of x may be neglected, show that

$$11\sin x - 6\cos x + 5 = A + Bx + Cx^2$$

and state the values of the constants A, B and C.   [E]

15 Show that $x = 0$ is a solution of the equation

$$\begin{vmatrix} x-1 & 4 & -1 \\ 1 & x+2 & 1 \\ 2x-4 & 4 & x-4 \end{vmatrix} = 0$$

and find the other two roots.   [E]

16 Show that the matrix

$$\begin{pmatrix} 4 & -3 & 0 \\ 1 & 3 & 2 \\ -1 & 0 & 1 \end{pmatrix}$$

has only one real eigenvalue and find a normalised eigenvector corresponding to this eigenvalue.   [E]

17 Express $\dfrac{\sqrt{(1+x)}}{1-x}$ as a series of ascending powers of x up to and including the term in x^2. Use this series to find, correct to 2 decimal places, the percentage change in the value of $\dfrac{a^{\frac{1}{2}}}{b}$ produced by an increase of 1% in the value of a and a decrease of 1% in the value of b.

18 A pyramid has a square base $OPQR$ and vertex S. Referred to O, the points P, Q, R and S have position vectors $\overrightarrow{OP} = 2\mathbf{i}$, $\overrightarrow{OQ} = 2\mathbf{i} + 2\mathbf{j}$, $\overrightarrow{OR} = 2\mathbf{j}$, $\overrightarrow{OS} = \mathbf{i} + \mathbf{j} + 4\mathbf{k}$.

(a) Express $\overrightarrow{PS}$ in terms of $\mathbf{i}$, $\mathbf{j}$ and $\mathbf{k}$.

(b) Show that the vector $-4\mathbf{j} + \mathbf{k}$ is perpendicular to OS and PS.

(c) Find, to the nearest degree, the acute angle between the line SQ and the plane OSP.   [E]

19
$$f(x) \equiv \frac{7 - 3\cos x - 6\sin x}{9 - 3\cos x - 8\sin x}$$

Prove that $\frac{1}{2} \leqslant f(x) \leqslant 1$. [E]

20 Given that $x > 0$, prove that
$$x > \sin x > x - \tfrac{1}{6}x^3$$ [E]

21 With respect to an origin O, the straight lines l_1 and l_2 have equations
$$l_1 : \mathbf{r} = p\mathbf{i} - 2\mathbf{j} + 2\mathbf{k} + \lambda(\mathbf{i} - \mathbf{k})$$
$$l_2 : \mathbf{r} = 3\mathbf{i} - \mathbf{j} + \mu(2\mathbf{i} + \mathbf{j} - 3\mathbf{k})$$

where λ and μ are scalar parameters and p is a scalar constant. The lines intersect at the point A.

(a) Find the coordinates of A and show that $p = 2$.

The plane Π passes through A and is perpendicular to l_2.

(b) Find a cartesian equation of Π.

(c) Find the acute angle between the plane Π and the line l_1, giving your answer in degrees to 1 decimal place. [E]

22 (a) Sketch on the same Argand diagram, the curves with equations $|z + 1|^2 = 12$ and $\arg(z - 1) = \frac{\pi}{4}$.

(b) Calculate the coordinates of the point of intersection of these two loci. [E]

23 (a) Find the angle between the vectors $\mathbf{i} - 2\mathbf{j} + 2\mathbf{k}$ and $-\mathbf{i} + 4\mathbf{j} + 8\mathbf{k}$.

Referred to an origin O, the points A and B are given by $\overrightarrow{OA} = \mathbf{i} - 2\mathbf{j} + 2\mathbf{k}$ and $\overrightarrow{OB} = -\mathbf{i} + 4\mathbf{j} + 8\mathbf{k}$.

(b) Find, in the form $\mathbf{r} = \mathbf{p} + \mathbf{q}t$ an equation of the line L passing through A and B.

The points P and Q lie on L. The point P has x-coordinate zero and the point Q has y-coordinate zero.

(c) Find PQ^2 to 1 decimal place.

24 Starting from the definitions of $\sinh x$ and $\cosh x$ in terms of e^x, prove that

(a) $\cosh^2 x + \sinh^2 x \equiv \cosh 2x$,

(b) $\cosh x + \cosh 3x \equiv 2\cosh x \cosh 2x$.

The function $\cosh x \cosh 2x$ is expanded in ascending powers of x.

(c) Find the terms of the expansion up to and including the term in x^4.

25 Given that $u_{n+1} = 2u_n + 1$, $u_1 = 1$, prove by induction that
$u_n = 2^n - 1$.

26 A transformation from the z-plane to the w-plane is given by
$$w = \frac{az + b}{z + c}$$
where a, b and c are real numbers.
The transformation maps $3 + 4i$ to $3 - 4i$ and also maps $z = 0$
to $w = 0$.
(a) Find the values of a, b and c.
(b) Find the other point which remains invariant under the
transformation.   [E]

27
$$\mathbf{A} = \begin{pmatrix} 1 & 0 & 1 \\ 3 & 1 & 1 \\ 4 & 2 & 7 \end{pmatrix}$$

(a) Without using a calculator, find $\mathbf{A}^{-1}$.
The transformation $T: \mathbb{R}^3 \to \mathbb{R}^3$ is represented by the matrix $\mathbf{A}$.
(b) Find the cartesian equations of the line which is mapped
by T onto the line
$$x = \frac{y}{4} = \frac{z}{3}$$ [E]

28 Find the expansion, in ascending powers of x, of $\ln\left(\frac{1+x}{1-x}\right)$,
giving the first three non-zero terms.
State the set of values of x for which the expansion is valid.
Find values of the positive constants a and b such that the
expansion of
$$2x(1 + ax)(1 + bx)^{-\frac{2}{3}}$$
in ascending powers of x is identical with that for $\ln\left(\frac{1+x}{1-x}\right)$
up to and including the term in x^3.   [E]

29 Given that $n \in \mathbb{Z}^+$, $\theta \in \mathbb{R}$ and
$$\mathbf{M} = \begin{pmatrix} \cosh^2\theta & \cosh^2\theta \\ -\sinh^2\theta & -\sinh^2\theta \end{pmatrix}$$
use induction to prove that $\mathbf{M}^n = \mathbf{M}$.   [E]

30 Find the perpendicular distance between the planes with
equations $2x + 2y + z - 6 = 0$, $2x + 2y + z - 10 = 0$. Find also
the area of the triangle whose vertices are $(2, 2, 2)$, $(1, 1, 2)$
and $(1, -1, 6)$.   [E]

31 Prove, by induction or otherwise, that for any positive integer n,

$$n^3 + 6n^2 + 8n$$

is divisible by 3. [E]

32 Prove by induction that if n is a positive integer

$$(\cos\theta + i\ \sin\theta)^n = \cos n\theta + i\ \sin n\theta$$

Evaluate $(1 + i)^n + (1 - i)^n$ when $n = 20$. [E]

33 (a) Obtain the first 4 non-zero terms of the binomial expansion in ascending powers of x of

$$(1 - x^2)^{-\frac{1}{2}}, \text{ given that } |x| < 1$$

(b) Show that, when $x = \frac{1}{3}$, $(1 - x^2)^{-\frac{1}{2}} = \frac{3}{4}\sqrt{2}$.

(c) Substitute $x = \frac{1}{3}$ into your expansion and hence obtain an approximation to $\sqrt{2}$, giving your answer to 5 decimal places. [E]

34 A transformation from the z-plane to the w-plane is given by

$$w = \frac{z - i}{z}$$

Show that under this transformation the line $\text{Im}(z) = \frac{1}{2}$ is mapped to the circle with equation $|w| = 1$.

Hence, or otherwise, find, in the form $w = \dfrac{az + b}{cz + d}$ where a, b, c and $d \in \mathbb{C}$, the transformation that maps the line $\text{Im}(z) = \frac{1}{2}$ to the circle centre $(3 - i)$ and radius 2. [E]

35 (a) Shade on an Argand diagram the region R for which
$$|z - 2 - 3i| < 1.$$

(c) Use de Moivre's theorem to prove that

$$\sin 7\theta = 7\sin\theta - 56\sin^3\theta + 112\sin^5\theta - 64\sin^7\theta$$

Deduce, or prove otherwise, that the only real solutions of the equation

$$\sin 7\theta = 7\sin\theta$$

are given by $\theta = n\pi$, where n is an integer. [E]

36 Find the first three non-zero terms of the expansion, in ascending powers of x, of $\ln(1 + xe^x)$. [E]

37
$$\mathbf{A} = \begin{pmatrix} 6 & 2 & -3 \\ 2 & 0 & 0 \\ -3 & 0 & 2 \end{pmatrix}$$

(a) Given that $\lambda = -1$ and $\lambda = 8$ are two eigenvalues of the matrix $\mathbf{A}$, find the third eigenvalue.

(b) Find the normalised eigenvector corresponding to the eigenvalue $\lambda = 8$.

Given that $\begin{pmatrix} \frac{1}{\sqrt{14}} \\ \frac{2}{\sqrt{14}} \\ \frac{3}{\sqrt{14}} \end{pmatrix}$ and $\begin{pmatrix} \frac{1}{\sqrt{6}} \\ -\frac{2}{\sqrt{6}} \\ \frac{1}{\sqrt{6}} \end{pmatrix}$ are eigenvectors

corresponding to the other two eigenvalues,

(c) write down a matrix $\mathbf{P}$ such that $\mathbf{P}^T\mathbf{A}\mathbf{P}$ is a diagonal matrix.

(d) State the elements of the matrix $\mathbf{P}^T\mathbf{A}\mathbf{P}$. [E]

38 The points $A(1, 7, -3)$, $B(1, 3, -1)$ and $C(5, 1, 2)$ have position vectors $\mathbf{a}$, $\mathbf{b}$ and $\mathbf{c}$ respectively relative to a fixed origin O.

(a) Find $\mathbf{b} \times \mathbf{c}$, and show that it is parallel to $\mathbf{i} - \mathbf{j} - 2\mathbf{k}$.

(b) Find $\mathbf{a} \cdot (\mathbf{b} \times \mathbf{c})$, and interpret your result geometrically. [E]

39 Given that $f(x) \equiv \ln(2x + \sqrt{(1 + 4x^2)})$,

(a) show that $f'(x) \equiv \dfrac{2}{\sqrt{(1 + 4x^2)}}$.

(b) Obtain the Maclaurin expansion for $f(x)$ in ascending powers of x, up to and including the term in x^3.

This series expansion is the same as the Maclaurin series expansion for $\sin(kx)$, up to and including the term in x^3.

(c) Write down the value of k.

40 Given that the first 2 non-zero terms in the expansion in ascending powers of x of

$$e^{ax} \cos 2x - \ln(1 + bx) - 1$$

are $7x^2$ and $-\frac{21}{2}x^3$, find the values of a and b. [E]

41 Given that $\mathbf{A} = \begin{pmatrix} 1 & 1 & 1 \\ 1 & 3 & 6 \\ 1 & 2 & 3 \end{pmatrix}$, find $\mathbf{A}^{-1}$.

Given that $\mathbf{Y} = \mathbf{A}\mathbf{X}$ and $\mathbf{Y} = \begin{pmatrix} 1 \\ 2 \\ 1 \end{pmatrix}$, hence, or otherwise,

find $\mathbf{X}$. [E]

42 Find the first three derivatives of $(1 + x)^2 \ln(1 + x)$. Hence, or otherwise, find the expansion of $(1 + x)^2 \ln(1 + x)$ in ascending powers of x up to and including the term in x^3 [E]

43 Given that $\mathbf{A} = \begin{pmatrix} 4 & -3 & 1 \\ 2 & 1 & -4 \\ 1 & 2 & -2 \end{pmatrix}$, find $\mathbf{A}^{-1}$. [E]

44 The point (x, y) is transformed to the point (x', y') by means of the transformation

$$\begin{pmatrix} x' \\ y' \end{pmatrix} = \begin{pmatrix} 3 & 0 \\ 0 & 4 \end{pmatrix} \begin{pmatrix} x \\ y \end{pmatrix} + \begin{pmatrix} 1 \\ 1 \end{pmatrix}$$

Find the image of the line $y = 2x$ under this transformation. [E]

45 (a) Use de Moivre's theorem to express $\cos 5\theta$ in terms of powers of $\cos \theta$.
(b) Solve the equation $z^3 = 8\mathrm{i}$, giving each of the roots in the form $r\mathrm{e}^{\mathrm{i}\theta}$, where $r > 0$ and $0 \leqslant \theta \leqslant 2\pi$. [E]

46 A transformation from the z-plane to the w-plane is given by

$$w = \frac{2z - 1}{z - 2}$$

Show that the circle with equation $|z| = 1$ is mapped to the circle with equation $|w| = 1$. [E]

47 Sketch the locus of points z in the Argand diagram which satisfy the equation
$$|z - 1 - \mathrm{i}| = 4$$
and find the maximum and minimum values of $|z|$ for points on this locus. [E]

48 A transformation from the z-plane to the w-plane is given by

$$w = z^2$$

Show that, when z describes the line $y = 1$, w describes a parabola and find the equation of this parabola. [E]

49 $$\frac{\mathrm{d}y}{\mathrm{d}x} = 2 + x + \sin y \text{ with } y = 0 \text{ at } x = 0$$

(a) Use the Taylor series method to obtain y as a series in ascending powers of x up to and including the term in x^3, and hence obtain an approximate value for y at $x = 0.1$.

(b) Use the approximation $\left(\dfrac{dy}{dx}\right)_0 \approx \dfrac{y_1 - y_{-1}}{2h}$, with $h = 0.1$, together with your value of y at $x = 0.1$, to obtain an approximate value for y at $x = 0.2$, giving your answer to 3 decimal places.   [E]

50 Given that $3u_{n+1} = 2u_n - 1$, $u_1 = 1$, prove by induction that $u_n = 3(\frac{2}{3})^n - 1$

51 The vertices O, A, B of a triangle in the Argand diagram are the points corresponding to the numbers 0, 1, $1 + i$ respectively. Show that, when the point z describes the perimeter of the triangle OAB, the locus of the point representing z^2 consists of part of the real axis, part of a parabola and part of the imaginary axis. Sketch this locus.   [E]

52 Given the differential equation

$$\frac{d^2y}{dx^2} + 3x\frac{dy}{dx} + 4y = 0$$

with $y = 1$ and $\dfrac{dy}{dx} = 1$ at $x = 0$,

(a) find y as a series in ascending powers of x up to and including the term in x^4.

(b) Use your series to obtain the value of y at $x = 0.1$, giving your answer to 2 decimal places.   [E]

53 The points $A(3, 0, 0)$, $B(0, 2, -1)$, and $C(2, 0, 1)$ have position vectors $\mathbf{a}$, $\mathbf{b}$ and $\mathbf{c}$ with respect to a fixed origin O. The line L has equation $(\mathbf{r} - \mathbf{a}) \times \mathbf{b} = \mathbf{0}$. The plane Π contains L and the point C.
(a) Find $\overrightarrow{AC} \times \overrightarrow{OB}$.
(b) Hence or otherwise show that an equation of Π is $2x + y + 2z = 6$.
(c) Find the perpendicular distance of Π from O.
The point R is the reflection of O in Π.
(d) Find the position vector of R.   [E]

54 Expand $\ln(1 + \sin x)$ in ascending powers of x up to and including the term in x^4.
Hence find an approximation for

$$\int_0^{\frac{\pi}{6}} \ln(1 + \sin x)\,dx$$

giving your answer to 3 decimal places.   [E]

55 (a) Given that $z = \cos\theta + i\sin\theta$, show that

$$z^n + \frac{1}{z^n} = 2\cos n\theta$$

where n is a positive integer.

(b) Hence show that $\cos^4\theta$ can be expressed in the form $p\cos 4\theta + q\cos 2\theta + r$, where p, q and r are constants, stating the values of p, q and r. [E]

56 (a) Express $z = \left(1 + i\sqrt{3}\right)$ in the form $r(\cos\theta + i\sin\theta)$, $r > 0, -\pi < \theta \leqslant \pi$.

(b) Hence, or otherwise, show that the two solutions of

$$w^2 = (1 + i\sqrt{3})^3$$

are $\left(2\sqrt{2}\right)i$ and $-\left(2\sqrt{2}\right)i$. [E]

57 Expand $(1 + x)^{\frac{1}{2}}$ in ascending powers of x up to and including the term in x^3. Given that the expansion of

$$\frac{(1 + x)^{\frac{1}{2}}}{1 - ax}$$

in ascending powers of x as far as the term in x^2 is $1 + bx^2$, find the values of a and b. [E]

58 By using the power series expansion for $\cos x$ and the power series expansion for $\ln(1 + x)$, find the series expansion for $\ln(\cos x)$ in ascending powers of x up to and including the term in x^4.

Hence, or otherwise, obtain the first two non-zero terms in the series expansion for $\ln(\sec x)$ in ascending powers of x. [E]

59 The lines L_1 and L_2 have equations $\mathbf{r} = \mathbf{a}_1 + s\mathbf{b}_1$ and $\mathbf{r} = \mathbf{a}_2 + t\mathbf{b}_2$ respectively, where

$$\mathbf{a}_1 = 3\mathbf{i} - 3\mathbf{j} - 2\mathbf{k}, \quad \mathbf{b}_1 = \mathbf{j} + 2\mathbf{k},$$
$$\mathbf{a}_2 = 8\mathbf{i} + 3\mathbf{j}, \qquad \mathbf{b}_2 = 5\mathbf{i} + 4\mathbf{j} - 2\mathbf{k}$$

(a) Verify that the point P with position vector $3\mathbf{i} - \mathbf{j} + 2\mathbf{k}$ lies on both L_1 and L_2.

(b) Find $\mathbf{b}_1 \times \mathbf{b}_2$.

(c) Find a cartesian equation of the plane containing L_1 and L_2.

The points with position vectors $\mathbf{a}_1$ and $\mathbf{a}_2$ are A_1 and A_2 respectively.

(d) By expressing $\overrightarrow{A_1P}$ and $\overrightarrow{A_2P}$ as multiples of $\mathbf{b}_1$ and $\mathbf{b}_2$ respectively, or otherwise, find the area of the triangle PA_1A_2. [E]

60 Prove by the method of mathematical induction, that, for $n \in \mathbb{Z}^+$,

$$\sum_{r=1}^{n} (r-1)(r+) = \frac{n}{1}) = \frac{n}{6}(n-1)(2n+5) \qquad \text{[E]}$$

61 The plane Π passes through $A(3, -5, -1)$, $B(-1, 5, 7)$ and $C(2, -3, 0)$.

(a) Find $\overrightarrow{AC} \times \overrightarrow{BC}$.

(b) Hence, or otherwise, find an equation, in the form $\mathbf{r} . \mathbf{n} = p$, of the plane Π.

(c) The perpendicular from the point $(2, 3, -2)$ to Π meets the plane at P. Find the coordinates of P. [E]

62 Prove that, for all positive integers, n,

$$7^{2n} + (2^{3n-3})(3^{n-1})$$

is divisible by 25. [E]

63 $\qquad \dfrac{\mathrm{d}y}{\mathrm{d}x} - x^2 - y^2 = 0$, with $y = 1$ at $x = 0$

Using the approximation $\left(\dfrac{\mathrm{d}y}{\mathrm{d}x}\right)_0 \approx \dfrac{y_1 - y_0}{h}$ twice, with step lengths of 0.1, find an approximate value for y at $x = 0.2$. [E]

64 $\qquad \dfrac{\mathrm{d}^2y}{\mathrm{d}x^2} - 4\dfrac{\mathrm{d}y}{\mathrm{d}x} + 3y^2 = 6$, with $y = 1$ and $\dfrac{\mathrm{d}y}{\mathrm{d}x} = 0$ at $x = 0$.

Use the Taylor series method to obtain y as a series in ascending powers of x up to and including the term in x^4. Hence find an approximate value for y at $x = 0.2$. [E]

65 (a) By using the series expansions of e^x and $\sin x$, or otherwise, show that the first three non-zero terms in the expansion of $\mathrm{e}^{\sin x}$ in ascending powers of x are

$$1 + x + \tfrac{1}{2}x^2$$

and that the coefficient of x^3 in this expansion is zero.

(b) Write down the first three non-zero terms in the expansion of $\mathrm{e}^{-\sin x}$ in ascending powers of x.

The coefficient of x^5 in the expansion of $\mathrm{e}^{\sin x}$ is $-\frac{1}{15}$.

(c) Find the coefficient of x^5 in the expansion of $\mathrm{e}^{-\sin x}$. [E]

66 Given that x is sufficiently small, use the approximations $\sin x \approx x$ and $\cos x \approx 1 - \tfrac{1}{2}x^2$ to show that

$$\frac{\cos x}{1 + \sin x} \approx 1 - x + \tfrac{1}{2}x^2$$

A student estimates the value of $\dfrac{\cos x}{1 + \sin x}$ when $x = 0.1$ by evaluating the approximation $1 - x + \tfrac{1}{2}x^2$ when $x = 0.1$. Find, to 3 decimal places, the percentage error made by the student.

[E]

67 The approximation $\left(\dfrac{dy}{dx}\right)_0 \approx \dfrac{y_1 - y_{-1}}{2h}$ is to be used to solve the differential equation

$$\frac{dy}{dx} + xy^2 = 3, \text{ with } y = y_0 \text{ at } x = x_0$$

using a step length of 0.1.

Show that application of the approximation gives

$$y_1 \approx (0.2)(3 - x_0 y_0{}^2) + y_{-1} \qquad\qquad \text{(I)}$$

Using the approximation

$$\left(\frac{dy}{dx}\right)_0 \approx \frac{y_0 - y_{-1}}{h} \text{ with } y_0 = 1 \text{ and } x_0 = 0$$

show that the value of y at $x = -0.1$ is 0.7.

Use (I) to find the value of y at $x = 0.3$, giving your answer to 3 decimal places.

[E]

68 A transformation T of the z-plane to the w-plane is given by

$$w = \frac{1 + 3z}{1 - z}, \quad z \neq 1$$

where $z = x + iy$ and $w = u + iv$.

The line $x = 0$ is mapped by T onto the curve C.

(a) Show that the equation of C is $(u + 1)^2 + v^2 = 4$.

(b) Sketch C in an Argand diagram.

[E]

69 A is a 3×3 matrix of the form $\begin{pmatrix} a & b & c \\ 0 & d & e \\ 0 & 0 & f \end{pmatrix}$ where a, b, c, d, e, f are real.

The transformation represented by $\mathbf{A}$ maps the points $\begin{pmatrix} 1 \\ 0 \\ 1 \end{pmatrix}$

and $\begin{pmatrix} 2 \\ 1 \\ 1 \end{pmatrix}$ onto the points $\begin{pmatrix} 2 \\ 3 \\ 4 \end{pmatrix}$ and $\begin{pmatrix} 1 \\ -1 \\ 4 \end{pmatrix}$ respectively.

(a) Given that $\det \mathbf{A} = 8$, find $\mathbf{A}$.

$$\mathbf{B} = \begin{pmatrix} 0 & -1 & 1 \\ 2 & -1 & 2 \\ 1 & 0 & 1 \end{pmatrix}, \mathbf{I} = \begin{pmatrix} 1 & 0 & 0 \\ 0 & 1 & 0 \\ 0 & 0 & 1 \end{pmatrix}$$

(b) Show that $\mathbf{B}^3 = \mathbf{I}$, and hence find $\mathbf{B}^{-1}$.

The matrix $\mathbf{C}$ satisfies the equation

$$\mathbf{B(C + I)} = \mathbf{(B + I)}^2$$

(c) Find p, q, r such that the image of $\begin{pmatrix} p \\ q \\ r \end{pmatrix}$ under the

transformation represented by $\mathbf{C}^{-1}\mathbf{B}^{-1}$ is $\begin{pmatrix} 1 \\ 1 \\ -3 \end{pmatrix}$. [E]

70 Given that

$$\mathrm{Im}\left(\frac{2z + 1}{iz + 1}\right) = -2$$

show that the locus of the point representing z in the Argand diagram is a straight line. [E]

71 Find an expansion for $\sec x$ in ascending powers of x as far as the term containing x^4. [E]

72 Sketch the circle with equation $|z - 3| = 2$ on an Argand diagram. State the greatest value of $|z|$ when $|z - 3| = 2$. [E]

73 By using

$$2\cos n\theta = e^{in\theta} + e^{-in\theta}$$

for suitable values of n, or otherwise, show that

$$2^5 \cos^6\theta = \cos 6\theta + 6\cos 4\theta + 15\cos 2\theta + 10$$

Hence, or otherwise, evaluate

$$\int_0^{\frac{\pi}{4}} 2^5 \cos^6\theta \, d\theta \qquad [\text{E}]$$

74 Show that the points in the Argand diagram representing the solutions of the equation $z^8 = 256$ lie at the vertices of a regular octagon. [E]

75 A sequence u_1, u_2, ... is defined by the relationships

$$u_n u_{n+1} + 2u_{n+1} - 4u_n - 3 = 0, \ u_1 = 2$$

Prove by induction that $u_n > 0$ and that $u_n < 3$, and hence prove that the sequence is increasing. [E]

76 Sketch on an Argand diagram the locus represented by the equation $|z - 1| = 1$. Shade on your diagram the region for which $|z - 1| < 1$ and $\frac{\pi}{6} < \arg z < \frac{\pi}{3}$. [E]

77 Use mathematical induction to prove that

$$\sum_{r=1}^{n} r(r+1) = \tfrac{1}{3}n(n+1)(n+2)$$ [E]

78 Prove, by induction, that $7^n + 5$ is divisible by 3 for all positive integers n. [E]

79 $\dfrac{d^2y}{dx^2} = 3x - 7y^2 + 5$, $y = 1$ and $\dfrac{dy}{dx} = \tfrac{1}{2}$ at $x = 0$.

(a) Use the Taylor series method to obtain y as a series in ascending powers of x up to and including the term in x^3, and hence obtain an approximate value for y at $x = 0.1$.

(b) Use the result

$$\left(\frac{d^2y}{dx^2}\right)_0 \approx \frac{y_1 - 2y_0 + y_{-1}}{h^2}$$

with $h = 0.1$, and your value of $y(0.1)$ obtained in (a), to find an approximate value for y at $x = 0.2$, giving your answer to 3 decimal places. [E]

80 The position vectors of the points P, Q and R relative to an origin O are respectively $2\mathbf{i} + 7\mathbf{k}$, $4\mathbf{j} + \mathbf{k}$, $\mathbf{i} + \mathbf{j} + \mathbf{k}$.

Find

(a) the area of $\triangle PQR$

(b) an equation of the plane PQR in the form $\mathbf{r}.\mathbf{n} = p$. [E]

81 (a) With the usual notation, show that

$$\left(\frac{dy}{dx}\right)_0 \approx \frac{y_1 - y_0}{h}$$

Given that

$$\frac{dy}{dx} = \frac{x^2 + y^2}{x + y}$$

and that $y = 3$ at $x = 2$, estimate the value of y at $x = 2.05$ and at $x = 2.1$, giving your answers to 3 significant figures.

(b) Find the solution, in ascending powers of x up to and including the term in x^3, of the differential equation

$$\frac{d^2y}{dx^2} - (x + 2)\frac{dy}{dx} + 3y = 0$$

given that, when $x = 0$, $y = 2$ and $\dfrac{dy}{dx} = 4$. [E]

82 Prove by induction, or otherwise, that

$$\sum_{r=1}^{n} r^3 = \tfrac{1}{4}n^2(n+1)^2$$

Hence show that the sum of the cubes of all the numbers between 100 and 200 which are divisible by 3 is a multiple of 4200. [E]

83 The position vectors of the points A, B, C are

$$\mathbf{i} - \mathbf{j} + 2\mathbf{k}, \quad 2\mathbf{i} + \mathbf{j} + 4\mathbf{k}, \quad 3\mathbf{i} + 4\mathbf{k},$$

respectively. Find

(a) the angle BAC, to the nearest degree

(b) the area of the triangle ABC

(c) a vector equation of the plane ABC

(d) the distance from the plane ABC to the point with position vector $3\mathbf{i} + 4\mathbf{j} + 5\mathbf{k}$. [E]

84

$$\mathbf{M} = \begin{pmatrix} 1 & 0 & 0 \\ x & 2 & 0 \\ 3 & 1 & 1 \end{pmatrix}$$

Find $\mathbf{M}^{-1}$ in terms of x. [E]

85 (a) If $z = 3 - 4\mathrm{i}$ and $w = 12 + 5\mathrm{i}$, express wz and $\dfrac{w}{z}$ in the form $a + b\mathrm{i}$, where a and b are real.

(b) Find the modulus and argument of $1 + \mathrm{i}\sqrt{3}$ and hence simplify $\left(1 + \mathrm{i}\sqrt{3}\right)^{10}$.

(c) Illustrate in an Argand diagram the set of complex numbers

$$A = \{z : |z + 6| = 2|z - 3\mathrm{i}|\}$$ [E]

86 (a) If $z = 2\mathrm{e}^{-\mathrm{i}\frac{\pi}{3}}$, express z, z^2 and $\dfrac{1}{z}$ in the form $x + y\mathrm{i}$ where x and y are real. State the modulus and argument of each of these complex numbers, giving the argument in each case as an angle θ such that $-\pi < \theta \leqslant \pi$. Represent the three numbers in an Argand diagram.

(b) By using de Moivre's theorem, or otherwise, show that

$$\sin 5\theta = 16\sin^5\theta - 20\sin^3\theta + 5\sin\theta$$ [E]

87 Given that $u_1 = 3$, $u_2 = 7$ and $u_n = 5u_{n-1} - 6u_{n-2}$ for $n \geqslant 3$, prove, by induction or otherwise, that $u_n = 2^n + 3^{n-1}$. [E]

88 Given that $x^2 + y^2 < 3x + 4y - 5$, show that $\dfrac{y}{x}$ lies between $\frac{1}{2}$ and $\frac{11}{2}$. [E]

89 Show geometrically or otherwise that for any two complex numbers z_1 and z_2.
$$|z_1 + z_2| \leqslant |z_1| + |z_2|$$
Prove by induction that the sum of the moduli of any finite number of complex numbers is not less than the modulus of their sum. [E]

90 Given that $\mathbf{M} = \begin{pmatrix} 1 & a \\ 0 & 1 \end{pmatrix}$, prove by induction that, when $n \in \mathbb{Z}^+$,
$$\mathbf{M}^n = \begin{pmatrix} 1 & na \\ 0 & 1 \end{pmatrix}$$ [E]

91 $\dfrac{dy}{dx} = y^2 + xy + x$, $y = 1$ at $x = 0$
(a) Use the Taylor series method to find y as a series in ascending powers of x up to and including the term in x^3.
(b) Use your series to find y at $x = 0.1$, giving your answer to 2 decimal places. [E]

92 Show that the equation of a plane can be expressed in the form
$$\mathbf{r} \cdot \mathbf{n} = p$$
Find an equation of the plane through the origin parallel to the lines $\mathbf{r} = 3\mathbf{i} + 3\mathbf{j} - \mathbf{k} + s(\mathbf{i} - \mathbf{j} - 2\mathbf{k})$ and $\mathbf{r} = 4\mathbf{i} + 5\mathbf{j} - 8\mathbf{k} + t(3\mathbf{i} + 7\mathbf{j} - 6\mathbf{k})$.
Show that one of the lines lies in the plane, and find the distance of the other line from the plane. [E]

93 $$\dfrac{d^2y}{dx^2} + x^2\dfrac{dy}{dx} + y = 0$$
with $y = 2$ at $x = 0$ and $\dfrac{dy}{dx} = 1$ at $x = 0$.

(a) Use the Taylor series method to express y as a polynomial in x up to and including the term in x^3.
(b) Show that at $x = 0$, $\dfrac{d^4y}{dx^4} = 0$. [E]

94 Prove, by induction or otherwise, that $81 \times 3^{2n} - 2^{2n}$ is divisible by 5 for all positive integers n.

95 A plane passes through $A(2, 2, 1)$, and is perpendicular to the line joining the origin to A. Write down a vector equation of the plane in the form $\mathbf{r} . \mathbf{n} = p$.   [E]

96 Use the method of induction to prove that
$(1 \times 4) + (2 \times 5) + (3 \times 6) + \ldots + n(n + 3) = \frac{1}{3}n(n + 1)(n + 5)$
where n is any positive integer.   [E]

97 The position vectors of the points P, Q, R relative to an origin O are respectively

$$3\mathbf{i} + 6\mathbf{k}, \quad 5\mathbf{j} + 3\mathbf{k}, \quad \mathbf{i} + \mathbf{k}$$

Find (a) the area of $\triangle PQR$
(b) an equation of the plane PQR in the form $\mathbf{r} . \mathbf{n} = p$, where $p \in \mathbb{R}$.

The transformation represented by the matrix $\mathbf{M}$ where

$$\mathbf{M} = \begin{pmatrix} 2 & 1 & 0 \\ 1 & -1 & 1 \\ 5 & 1 & 0 \end{pmatrix}$$

maps the points A, B, C to the points P, Q, R respectively.
(c) Find det $\mathbf{M}$ and hence, or otherwise, find the position vectors of the points A, B and C.   [E]

98 Show by induction, or otherwise, that, if n is an integer and $n > 1$, $7^n - 6n - 1$ is divisible by 36, and $5^n - 4n - 1$ is divisible by 16.
Hence, or otherwise, show that $7^n - 5^n - 2n$ is divisible by 4. [E]

99 Planes P_1 and P_2 have equations $\mathbf{r} . (\mathbf{i} - \mathbf{j} + 2\mathbf{k}) = 5$ and $3x - z = 2$ respectively.
Find the size of the acute angle between P_1 and P_2 giving your answer to the nearest 0.1 degree.   [E]

100 A sequence of numbers is defined by $u_{n+1} = au_n + b$, where a, b are constants and $a \neq 0$, $a \neq 1$.
Prove by induction that

$$u_n + \frac{b}{a - 1} = a^{n-1}\left(u_1 + \frac{b}{a - 1}\right) \qquad \text{[E]}$$

101 Find a vector equation of the plane P passing through the points A, B and C with position vectors $\begin{pmatrix} 1 \\ -1 \\ 2 \end{pmatrix}$, $\begin{pmatrix} 2 \\ 1 \\ 1 \end{pmatrix}$ and $\begin{pmatrix} 1 \\ -2 \\ 2 \end{pmatrix}$ respectively.

Find the area of $\triangle\,ABC$.

Find also the distance of Π from the point D whose position vector is $\begin{pmatrix} 3 \\ 1 \\ 1 \end{pmatrix}$.   [E]

102 Use the method of induction to prove that

$$\sum_{r=1}^{n} 3^{r-1} = \frac{3^n - 1}{2}, \ n \in \mathbb{Z}^+ \qquad \text{[E]}$$

103 The position vectors of the points A, B, C and D relative to a fixed origin O, are $(-\mathbf{j} + 2\mathbf{k})$, $(\mathbf{i} - 3\mathbf{j} + 5\mathbf{k})$, $(2\mathbf{i} - 2\mathbf{j} + 7\mathbf{k})$ and $(\mathbf{j} + 2\mathbf{k})$ respectively.

(a) Find $\mathbf{p} = \overrightarrow{AB} \times \overrightarrow{CD}$.

(b) Calculate $\overrightarrow{AC} \cdot \mathbf{p}$.

Hence determine the shortest distance between the line containing AB and the line containing CD.   [E]

104 The points $A(2, 0, -1)$ and $B(4, 3, 1)$ have position vectors $\mathbf{a}$ and $\mathbf{b}$ with respect to a fixed origin O.

(a) Find $\mathbf{a} \times \mathbf{b}$.

The plane Π_1 contains the points O, A and B.

(b) Verify that an equation of Π_1 is $x - 2y + 2z = 0$.

The plane Π_2 has equation $\mathbf{r} \cdot \mathbf{n} = d$ where $\mathbf{n} = 3\mathbf{i} + \mathbf{j} - \mathbf{k}$ and d is a constant. Given that B lies on Π_2,

(c) find the value of d.

The planes Π_1 and Π_2 intersect in the line L.

(d) Find an equation of L in the form $\mathbf{r} = \mathbf{p} + t\mathbf{q}$, where t is a parameter.

(e) Find the position vector of the point X on L where OX is perpendicular to L.   [E]

105 Show that 9 is one eigenvalue of the matrix **A**, where

$$\mathbf{A} = \begin{pmatrix} 11 & 2 & 8 \\ 2 & 2 & -10 \\ 8 & -10 & 5 \end{pmatrix}$$

and find the other two eigenvalues.
Find normalised eigenvectors corresponding to each of the three eigenvalues. [E]

106 **M** is the matrix $\begin{pmatrix} 1 & 0 & 1 \\ 0 & 2 & 0 \\ 4 & 3 & 1 \end{pmatrix}$.

(a) Find the eigenvalues and corresponding eigenvectors for the matrix **M**.
(b) A transformation $T : \mathbb{R}^3 \to \mathbb{R}^3$ is represented by the matrix **M**. Find cartesian equations of the image of the line with equations

$$\frac{x}{2} = y = \frac{z}{-1}$$

under this transformation. [E]

107 Find each of the roots of the equation $z^5 - 1 = 0$ in the form $r(\cos\theta + i\sin\theta)$ where $r > 0$ and $-\pi < \theta \leqslant \pi$.
(a) Given that α is the complex root of this equation with the smallest positive argument, show that the roots of $z^5 - 1 = 0$ can be written as $1, \alpha, \alpha^2, \alpha^3, \alpha^4$.
(b) Show that $\alpha^4 = \alpha^*$ and hence, or otherwise, obtain $z^5 - 1$ as a product of real linear and quadratic factors giving the coefficients in terms of integers and cosines.
(c) Show also that $z^5 - 1 = (z - 1)(z^4 + z^3 + z^2 + z + 1)$ and hence, or otherwise, find $\cos\left(\frac{2}{5}\pi\right)$, giving your answer in terms of surds. [E]

108 Show that $y = 1 + \int_0^x e^{t^2}\,dt$ is the solution of the differential equation $\dfrac{dy}{dx} = e^{x^2}$ for which $y = 1$ at $x = 0$.
Working throughout to 3 decimal places, obtain an estimate for y at $x = 0.2$ using the Taylor series method to obtain a solution in powers of x as far as the term in x^3. [E]

109
$$A = \begin{pmatrix} 5 & -4 & 2 \\ 0 & 3 & 0 \\ -2 & -5 & 10 \end{pmatrix}$$

(a) Show that 3 is an eigenvalue of **A** and find the other two eigenvalues.

(b) Show that $\begin{pmatrix} 1 \\ 1 \\ 1 \end{pmatrix}$ is an eigenvector of **A** corresponding to the eigenvalue 3.

(c) For each of the other eigenvalues, find a corresponding eigenvector. [E]

110
$$A = \begin{pmatrix} 2 & -2 & 0 \\ -2 & 1 & 2 \\ 0 & 2 & 5 \end{pmatrix}$$

(a) Show that $\begin{pmatrix} 2 \\ 3 \\ -1 \end{pmatrix}$ and $\begin{pmatrix} 2 \\ -1 \\ 1 \end{pmatrix}$ are eigenvectors of **A** giving their corresponding eigenvalues.

(b) Given that 6 is the third eigenvalue of **A**, find a corresponding eigenvector.

(c) Hence write down a matrix **P** such that $\mathbf{P}^{-1}\mathbf{AP}$ is a diagonal matrix. [E]

111 The transformation T is represented by the matrix **M**, where

$$M = \begin{pmatrix} 0 & 1 & 0 \\ 0 & 0 & 7 \\ 1 & 2 & a \end{pmatrix}$$

(a) Find $\mathbf{M}^{-1}$ in terms of a.

The position vector of the point P, relative to an origin O, is $\mathbf{i} - 7\mathbf{j} + 3\mathbf{k}$. Given that, when $a = 2$, T maps the point Q onto the point P,

(b) find the position vector of the point Q. [E]

112 Represent on one Argand diagram
(a) the three roots of the equation $z^3 + 64 = 0$
(b) the six roots of the equation $z^6 + 64 = 0$. [E]

113 Given that y satisfies the differential equation

$$\frac{\mathrm{d}^2 y}{\mathrm{d}x^2} + 20\frac{\mathrm{d}y}{\mathrm{d}x} - y^2 = x$$

use the approximations

$$\left(\frac{\mathrm{d}y}{\mathrm{d}x}\right)_0 \approx \frac{y_1 - y_{-1}}{2h}, \quad \left(\frac{\mathrm{d}^2 y}{\mathrm{d}x^2}\right)_0 \approx \frac{y_1 - 2y_0 + y_{-1}}{h^2}$$

to find the value of y at $x = 0.3$, given also that $y = 1$ at $x = 0$ and $y = 2$ at $x = 0.1$. Give your final answer to 3 decimal places.

[E]

114 Obtain the series solution in ascending powers of x, up to and including the term in x^3, of the differential equation

$$\frac{\mathrm{d}^2 y}{\mathrm{d}x^2} + y\frac{\mathrm{d}y}{\mathrm{d}x} - 4y = 0$$

given that $y = 3$ and $\dfrac{\mathrm{d}y}{\mathrm{d}x} = 2$ at $x = 0$. [E]

115 Find the solution, in ascending powers of x up to and including the term in x^3, of the differential equation

$$\frac{\mathrm{d}^2 y}{\mathrm{d}x^2} - (x+2)\frac{\mathrm{d}y}{\mathrm{d}x} + 3y = 0$$

given that, at $x = 0$, $y = 2$ and $\dfrac{\mathrm{d}y}{\mathrm{d}x} = 4$. [E]

116 Show that the lines with equations

$$\mathbf{r} = (-2\mathbf{i} + 5\mathbf{j} - 11\mathbf{k}) + s(3\mathbf{i} + \mathbf{j} + 3\mathbf{k})$$
$$\mathbf{r} = (8\mathbf{i} + 9\mathbf{j}) + t(4\mathbf{i} + 2\mathbf{j} + 5\mathbf{k})$$

intersect and find the position vector of their common point, P. Show also that the plane α containing these lines is parallel to the plane β with equation

$$\mathbf{r} \cdot (\mathbf{i} + 3\mathbf{j} - 2\mathbf{k}) = 7$$

Find

(a) the position vector of N, the foot of the perpendicular from P to the plane β

(b) the distance between the planes α and β. [E]

Examination style paper

FP3

1. Given that n is a positive integer, greater than 2, use the method of mathematical induction to prove that $2^n > 2n$. **(6 marks)**

2. The z-plane is mapped on to the w-plane by the transformation $w = z + \dfrac{1}{z}$. Given that z lies on the circle $|z| = 1$, show that w lies on an interval of the real axis. Identify this interval precisely.
    (8 marks)

3. Given that
    $$\sin x = x - \frac{1}{3!}x^3 + \frac{1}{5!}x^5 - \cdots,$$
    $$\cos x = 1 - \frac{1}{2!}x^2 + \frac{1}{4!}x^4 - \cdots,$$

    prove that
    $$\tan x = 1 + \frac{1}{3}x^3 + \frac{2}{15}x^5 + \cdots \qquad \textbf{(4 marks)}$$

    Hence determine the series expansion for $e^{\tan x}$ in ascending powers of x up to and including the term in x^4. **(5 marks)**

4. Find in the form $k\,e^{i\alpha}$, where $k \in \mathbb{R}^+$ and $-\pi < \alpha \leqslant \pi$ all the roots of the equation
    $$z^4 = 8 + 8i\sqrt{3} \qquad \textbf{(9 marks)}$$

    Display the roots on an Argand diagram. **(2 marks)**

5. $$\frac{d^2y}{dx^2} + x\frac{dy}{dx} + y = 0, \quad y = 1 \quad \text{and} \quad \frac{dy}{dx} = 0 \quad \text{at} \quad x = 0.$$

    Use the approximations
    $$\left(\frac{dy}{dx}\right)_0 \approx \frac{y_1 - y_{-1}}{2h} \quad \text{and} \quad \left(\frac{d^2y}{dx^2}\right)_0 \approx \frac{y_1 - 2y_0 + y_{-1}}{h^2}$$

    with a step width h of 0.1 to find an approximate value of y at $x = 0.2$, giving your answer to 4 decimal places. **(12 marks)**

6.
$$A = \begin{pmatrix} 2 & -2 & 3 \\ 1 & 1 & 1 \\ 1 & 3 & -1 \end{pmatrix}$$

    (*a*) Find the eigenvalues of the matrix **A**.

    (*b*) For each eigenvalue, find a corresponding eigenvector.

(13 marks)

7.   The plane P which has equation

$$\mathbf{r} \cdot (2\mathbf{i} - 3\mathbf{j} + \mathbf{k}) = 7$$

meets the line *l* which has equation

$$\mathbf{r} = (3\mathbf{i} + 5\mathbf{j} + \mathbf{k}) + t(2\mathbf{i} + 6\mathbf{j} - \mathbf{k})$$

in the point *P*.

    (*a*) Find the coordinates of *P*.

    (*b*) Find the shortest distance of P from the origin *O*.

       The line *m* passes through the point *P* and the point *Q*, where $\overrightarrow{OQ} = 4\mathbf{i} + 7\mathbf{j}$.

    (*c*) Determine a vector which is perpendicular to both of the lines *l* and *m*.

(16 marks)

Answers

Edexcel accepts no responsibility whatsoever for the accuracy or method of working in the answers given for examination questions.

Exercise 1A

6 $x + \frac{1}{3}x^3 + \frac{2}{15}x^5 + \dots$

7 $x^2 - \frac{1}{3}x^4 + \frac{2}{45}x^6 - \dots$

8 $2x + \frac{2}{3}x^3 + \frac{2}{5}x^5 + \dots$

9 $1 - x^2 - \frac{1}{2}x^4 - \dots$

10 $1 + x - \frac{1}{3}x^3 + \dots$

11 $x + \frac{1}{6}x^3 + \frac{1}{120}x^5 + \dots$

12 $x + \frac{1}{3}x^3 - \frac{1}{30}x^5 + \dots$

13 $1 + x^2 + \frac{1}{3}x^4 + \dots$

14 $x - x^3 + x^5 - \dots$

15 $1 + x - \frac{3}{2}x^2 + \dots$

Exercise 1B

1 $A = 2, B = -\frac{7}{6}$

2 $C = 1, D = \frac{1}{3}$

3 $\frac{1}{4}\sqrt{3}$

7 1.43

8 $A = 0$

9 (a) $-\frac{1}{2}$ (b) $-\frac{1}{2}$

Exercise 1C

2 $1 + \frac{1}{2}(x - 1) - \frac{1}{8}(x - 1)^2 + \frac{1}{16}(x - 1)^3$

3 (a) $e^h\left[1 + (x - h) + \frac{1}{2}(x - h)^2\right]$

    (b) $\frac{1}{h} - \frac{1}{h^2}(x - h) + \frac{1}{h^3}(x - h)^2$

4 $(x - 1) + \frac{3}{2}(x - 1)^2 + \frac{1}{3}(x - 1)^3 + \dots$

5 $\frac{1}{2} + \frac{\sqrt{3}}{2}x - \frac{1}{4}x^2 - \frac{\sqrt{3}}{12}x^3 - \dots$

6 $\frac{1}{2} - \frac{\sqrt{3}}{2}x - \frac{1}{4}x^2 + \frac{\sqrt{3}}{12}x^3 - \dots$

7 $y = 1 - x + x^2 - \frac{2}{3}x^3$

8 $y = 2 + 2x + 2x^2 + \frac{5}{3}x^3$

9 $y = 1 + \frac{1}{2}(\cosh 1)x^2$

10 $y = 1 + x + \frac{1}{2}x^2$

11 $y = 1 + \frac{1}{2}x^2 + \frac{1}{6}x^3$

12 $y = 1 - \frac{1}{2}x^2$

13 $y = 1 - 2x + 2x^2 - \frac{5}{3}x^3$

14 $y = \frac{\pi}{4} + \frac{1}{2}x - \frac{1}{4}x^2 + \frac{1}{12}x^3$

15 $y = 1 + x - x^2 - \frac{1}{3}x^3$

16 $y = 1 + (x - 1) + (x - 1)^2 + \frac{2}{3}(x - 1)^3$

17 $y = (x - 1) + \frac{1}{2}(x - 1)^2$

18 $y = 1 + x + \frac{1}{2}x^2 + \frac{1}{2}x^3$

19 $0.009\,005\,2$

20 $0.198\,669$

21 $1 - \frac{1}{3}x^2 - \frac{1}{9}x^4;\ 0.3927$

Exercise 2A

1 (a) (i) $5\left(\cos \frac{\pi}{2} + i \sin \frac{\pi}{2}\right)$

      (ii) $5e^{\frac{i\pi}{2}}$

  (b) (i) $7(\cos 0 + i \sin 0)$

      (ii) $7e^0$

  (c) (i) $3\left[\cos\left(-\frac{\pi}{2}\right) + i \sin\left(-\frac{\pi}{2}\right)\right]$

      (ii) $3e^{-\frac{i\pi}{2}}$

  (d) (i) $6(\cos \pi + i \sin \pi)$

      (ii) $6e^{i\pi}$

  (e) (i) $2\left(\cos \frac{\pi}{3} + i \sin \frac{\pi}{3}\right)$

      (ii) $2e^{\frac{i\pi}{3}}$

  (f) (i) $6\left[\cos\left(-\frac{\pi}{6}\right) + i \sin\left(-\frac{\pi}{6}\right)\right]$

      (ii) $6e^{-\frac{i\pi}{6}}$

  (g) (i) $5(\cos 2.21 + i \sin 2.21)$

      (ii) $5e^{2.21i}$

  (h) (i) $\sqrt{2}\left[\cos\left(-\frac{\pi}{4}\right) + i \sin\left(-\frac{\pi}{4}\right)\right]$

      (ii) $\sqrt{2}e^{-\frac{i\pi}{4}}$

  (i) (i) $10[\cos(-0.927) + i \sin(-0.927)]$

      (ii) $10e^{-0.927i}$

  (j) (i) $\cos \frac{\pi}{3} + i \sin \frac{\pi}{3}$

      (ii) $e^{\frac{i\pi}{3}}$

(k) (i) $4(\cos\frac{\pi}{6} + i\sin\frac{\pi}{6})$

    (ii) $4e^{\frac{i\pi}{6}}$

(l) (i) $\frac{\sqrt{221}}{17}[\cos(-1.91) + i\sin(-1.91)]$

    (ii) $\frac{\sqrt{221}}{17}e^{-1.91i}$

2 (a) $\frac{3}{2}(1 + i\sqrt{3})$ (b) $-\frac{5}{\sqrt{2}} + \frac{5i}{\sqrt{2}}$

  (c) $3\sqrt{3} - 3i$ (d) $4i$

  (e) -10 (f) $-\frac{9}{2}(\sqrt{3} + i)$

  (g) $\frac{7}{3\sqrt{2}}(1 + i)$ (h) $\frac{3}{\sqrt{2}}(-1 + i)$

  (i) $4(-\sqrt{3} + i)$ (j) $\frac{2}{3}(\sqrt{3} + i)$

3 $e^{-\pi i} = -1$

4 $\sinh z \cosh w + \cosh z \sinh w$

5 $\dfrac{\tanh z + \tanh w}{1 + \tanh z \tanh w}$

6 $1 - \tanh^2 z$

7 $\frac{1}{10}e^x(\sin 3x - 3\cos 3x) + C$

8 $i\ln(3 \pm \sqrt{8})$

9 $\dfrac{\tanh x \sec^2 y}{1 + \tanh^2 x \tan^2 y}$, $\dfrac{i\tan y \operatorname{sech}^2 x}{1 + \tanh^2 x \tan^2 y}$

10 (a) $2m\pi \pm i\operatorname{arcosh} 4$

  (b) $(4m + 1)\frac{\pi}{2} \pm i\operatorname{arcosh} 2$

  (c) $n\pi + i\operatorname{arsinh}[(-1)^n]$

Exercise 2B

1 (a) $\cos 4\pi + i\sin 4\pi = 1$

  (b) $\cos\frac{8\pi}{12} + i\sin\frac{8\pi}{12} = -\frac{1}{2} + i\frac{\sqrt{3}}{2}$

  (c) $\cos(-\frac{\pi}{3}) + i\sin(-\frac{\pi}{3}) = \frac{1}{2} - i\frac{\sqrt{3}}{2}$

  (d) $\cos(-\frac{\pi}{6}) + i\sin(-\frac{\pi}{6}) = \frac{\sqrt{3}}{2} - \frac{1}{2}i$

2 $4[\cos(-\frac{\pi}{3}) + i\sin(-\frac{\pi}{3})]; 4^8(-\frac{1}{2} - i\frac{\sqrt{3}}{2});$

   $\frac{1}{4^5}(\frac{1}{2} - i\frac{\sqrt{3}}{2})$

3 $\sqrt{2}[\cos(-\frac{\pi}{4}) + i\sin(-\frac{\pi}{4})]; -4;$

   $\frac{1}{16}(1 - i)$

4 $(\cos\frac{\pi}{9} + i\sin\frac{\pi}{9})^9 = \cos\pi + i\sin\pi = -1$

5 $16\sin^5\theta - 20\sin^3\theta + 5\sin\theta$

6 $3\sin\theta - 4\sin^3\theta$

7 $-7\cos\theta + 56\cos^3\theta - 112\cos^5\theta + 64\cos^7\theta$

8 $7\sin\theta - 56\sin^3\theta + 112\sin^5\theta - 64\sin^7\theta$

9 $\dfrac{3\tan\theta - \tan^3\theta}{1 - 3\tan^2\theta}$

10 $\dfrac{5\tan\theta - 10\tan^3\theta + \tan^5\theta}{1 - 10\tan^2\theta + 5\tan^4\theta}$

11 (a) $\frac{1}{16}(\cos 5\theta + 5\cos 3\theta + 10\cos\theta)$

  (b) $\frac{1}{32}(\cos 6\theta + 6\cos 4\theta + 15\cos 2\theta + 10)$

  (c) $\frac{1}{64}(\cos 7\theta + 7\cos 5\theta + 21\cos 3\theta$
               $+ 35\cos\theta)$

12 (a) $\frac{1}{8}(\cos 4\theta - 4\cos 2\theta + 3)$

  (b) $\frac{1}{16}(\sin 5\theta - 5\sin 3\theta + 10\sin\theta)$

  (c) $-\frac{1}{64}(\sin 7\theta - 7\sin 5\theta + 21\sin 3\theta$
                $- 35\sin\theta)$

13 (a) $\frac{1}{8}(\frac{1}{4}\sin 4\theta - 2\sin 2\theta + 3\theta) + C$

  (b) $\frac{1}{32}(\frac{1}{6}\sin 6\theta + \frac{3}{2}\sin 4\theta$
          $+ \frac{15}{2}\sin 2\theta + 10\theta) + C$

  (c) $\frac{1}{192}\sin 6\theta + \frac{5}{64}\sin 4\theta - \frac{1}{64}\sin 2\theta$
         $+ \frac{\theta}{16} + C$

14 (a) $e^{i(\frac{2k\pi}{3} + \frac{\pi}{6})}$, $k = 0, 1, 2$

  (b) $e^{i(\frac{2k\pi}{3} + \frac{\pi}{3})}$, $k = 0, 1, 2$

  (c) $\sqrt[3]{(13)}e^{i(\frac{2k\pi}{3} + \frac{\alpha}{3})}$ where $\alpha = 1.965$

  (d) $e^{i(\frac{2k\pi}{3} + \frac{\pi}{6})}$, $k = 0, 1, 2$

15 $z = e^{(\frac{2k\pi}{5} + \frac{\pi}{5})i}$, $k = 0, 1, 2, 3, 4$

16 $\sqrt[3]{75}\,e^{i(\frac{1.287 + 2k\pi}{3})}$, $k = 0, 1, 2$

17 $1, -1, i, -i$

18 $\dfrac{2 \pm i\sqrt{5}}{3}, -\dfrac{1 \pm i\sqrt{3}}{2}$

20 $2[\cos(-\frac{\pi}{6}) + i\sin(-\frac{\pi}{6})]$, 2^9

21 $0, -\frac{3}{2} + \frac{\sqrt{3}}{2}i, -\frac{3}{2} - \frac{\sqrt{3}}{2}i$

23 $\cos\frac{\pi}{12} + i\sin\frac{\pi}{12}, \cos\frac{3\pi}{4} + i\sin\frac{3\pi}{4}$

   $\cos(-\frac{7\pi}{12}) + i\sin(-\frac{7\pi}{12}),$

   $\cos(-\frac{\pi}{12}) + i\sin(-\frac{\pi}{12})$

   $\cos(-\frac{3\pi}{4}) + i\sin(-\frac{3\pi}{4}), \cos\frac{7\pi}{12} + i\sin\frac{7\pi}{12}$

24 $2e^{-\frac{5\pi}{6}i}, 2e^{-\frac{\pi}{3}i}, 2e^{\frac{\pi}{6}i}, 2e^{\frac{2\pi}{3}i}$

Exercise 2C

6 $3 + \sqrt{3}i, 3 - \sqrt{3}i$

7 $\frac{3}{2} + \frac{\sqrt{3}}{2}i; \frac{\pi}{6}; \frac{2\pi}{3}$ **8** $2 + 3i$

10 $\operatorname{Re}\left(z + \dfrac{1}{z}\right) = x + \dfrac{x}{x^2 + y^2}$

     $\operatorname{Im}\left(z + \dfrac{1}{z}\right) = y - \dfrac{y}{x^2 + y^2}$

11 centre $\left(-\frac{5}{4}, 0\right)$, radius $\frac{3}{4}$

12 $\frac{10}{9} \pm \frac{4\sqrt{14}}{9}i$

13 $(x + 1)^2 + (y - 2)^2 = 20$

14 (b) centre $(0, 1)$, radius $\sqrt{2}$

Exercise 2D

7 perpendicular bisector of the line joining $(0, 0)$ to $(9, 0)$

8 perpendicular bisector of the line joining $(0, -2)$ to $(-1, 0)$

9 circle, centre $(3 + i)$, radius 8

10 (a) w moves in circle, centre O, radius 2; starting at $(2, 0)$ and moving anticlockwise it completes 2 circles for each complete circle that z covers.

(b) As z moves anticlockwise round circle, centre $(0, 0)$, radius 1, w moves round same circle clockwise, passing through starting point at same time.

(c) As z moves from $(1, 0)$ in anticlockwise direction round circle, centre $(0, 0)$, radius 1, w moves from $(-1, 0)$ anticlockwise round circle, centre $\left(-\frac{2}{3}, 0\right)$, radius $\frac{1}{3}$.

12 $u^2 + v^2 = 1$, where $u = \text{Re}(w)$ and $v = \text{Im}(w)$.

13 (a) points inside circle, centre $(3, 4)$, radius 1

(b) points outside circle, centre $(0, 0)$, radius 1

(c) points inside circle, centre $(0, 0)$, radius 1

14 $u = \dfrac{2x}{x^2 + (y + 1)^2}$, $v = \dfrac{x^2 + y^2 - 1}{x^2 + (y + 1)^2}$

17 $a = 2$, $b = 3$, $c = 1$

18 $\frac{1}{2}(\sqrt{7} + i)$, $\frac{1}{2}(-\sqrt{7} + i)$

Exercise 3A

2 (a) $T^{-1} : \begin{pmatrix} x \\ y \end{pmatrix} \mapsto \begin{pmatrix} x \\ \frac{1}{2}y \end{pmatrix}$

(b) $T^{-1} : \begin{pmatrix} x \\ y \end{pmatrix} \mapsto \begin{pmatrix} y \\ x \end{pmatrix}$

(c) $T^{-1} : \begin{pmatrix} x \\ y \end{pmatrix} \mapsto \begin{pmatrix} x \\ -y \end{pmatrix}$

(d) T^{-1} does not exist

(e) $T^{-1} : \begin{pmatrix} x \\ y \end{pmatrix} \mapsto \begin{pmatrix} \dfrac{2y - x}{5} \\ \dfrac{3x - y}{5} \end{pmatrix}$

(f) T^{-1} does not exist

(g) $T^{-1} : \begin{pmatrix} x \\ y \\ z \end{pmatrix} \mapsto \begin{pmatrix} x \\ y - x \\ z - y + x \end{pmatrix}$

(h) T^{-1} does not exist

(i) $T^{-1} : \begin{pmatrix} x \\ y \\ z \end{pmatrix} \mapsto \begin{pmatrix} \frac{1}{2}(3x - y - z) \\ \frac{1}{2}(y + z - x) \\ y - x \end{pmatrix}$

(j) T^{-1} does not exist

Exercise 3B

1 (a) $\begin{pmatrix} -7 & 7 \\ 10 & 3 \end{pmatrix}$ (b) $\begin{pmatrix} -16 & 8 \\ -4 & 1 \end{pmatrix}$

(c) $\begin{pmatrix} 19 & -10 \\ -9 & 13 \end{pmatrix}$ (d) $\begin{pmatrix} -12 & 12 \\ -3 & 9 \end{pmatrix}$

(e) $\begin{pmatrix} 32 & 4 & -23 \\ -11 & -6 & 48 \\ 6 & -20 & 45 \end{pmatrix}$

(f) $\begin{pmatrix} 67 & -11 & -26 \\ -3 & -10 & -5 \\ 31 & 12 & -31 \end{pmatrix}$

2 (a) $\begin{pmatrix} -2 & 10 \\ 3 & -8 \end{pmatrix}$ (b) $\begin{pmatrix} -14 & 7 \\ -6 & 4 \end{pmatrix}$

(c) $\begin{pmatrix} 8 \\ 7 \\ -6 \end{pmatrix}$ (d) $\begin{pmatrix} 8 & -9 & -22 \\ 6 & 0 & -30 \end{pmatrix}$

(e) $\begin{pmatrix} -8 & -3 & 28 \\ 29 & 7 & 11 \\ -2 & -2 & 4 \end{pmatrix}$

(f) $\begin{pmatrix} 20 & 11 & 1 \\ 32 & -14 & 30 \\ -22 & -24 & -18 \end{pmatrix}$

(g) $\begin{pmatrix} -15 & -1 & -8 \\ -14 & 5 & 7 \\ 37 & 21 & -48 \end{pmatrix}$

(h) $\begin{pmatrix} -9 & 39 & 32 \\ -9 & -11 & 16 \\ 13 & -8 & 8 \end{pmatrix}$

(i) $\begin{pmatrix} 10 & 29 & -31 \\ -2 & -10 & 2 \\ 5 & 4 & 3 \end{pmatrix}$

3 (a) 1   (b) -2 (c) -5 (d) 2

(e) -9 (f) 0   (g) 29   (h) 2

(i) -36 (j) -29 (k) 0   (l) 0

(m) 0   (n) -4 (o) 6   (p) -14

4 (a) $\frac{1}{13}\begin{pmatrix} 4 & -5 \\ 1 & 2 \end{pmatrix}$ (b) $-\frac{1}{19}\begin{pmatrix} 7 & -2 \\ 1 & -3 \end{pmatrix}$

(c) $-\frac{1}{5}\begin{pmatrix} -4 & 3 \\ -1 & 2 \end{pmatrix}$

(d) $\frac{1}{3}\begin{pmatrix} 2 & -1 \\ 3 & 0 \end{pmatrix}$

(e) $-\frac{1}{129}\begin{pmatrix} 22 & -7 \\ -9 & -3 \end{pmatrix}$

(f) $-\frac{1}{3}\begin{pmatrix} 0 & -3 & -2 \\ -3 & 6 & -1 \\ 0 & 0 & -1 \end{pmatrix}$

(g) $\frac{1}{6}\begin{pmatrix} -10 & -13 & -4 \\ -2 & -5 & -2 \\ 0 & -3 & 0 \end{pmatrix}$

(h) $-\frac{1}{12}\begin{pmatrix} 6 & -11 & 1 \\ 6 & -19 & 5 \\ -6 & 9 & -3 \end{pmatrix}$

(i) $\frac{1}{17}\begin{pmatrix} -4 & 6 & -1 \\ 11 & -8 & -10 \\ 6 & -9 & -7 \end{pmatrix}$

(j) $\frac{1}{5}\begin{pmatrix} 2 & -1 & 1 \\ 5 & -10 & 0 \\ 3 & -4 & -1 \end{pmatrix}$

(k) $\frac{1}{3}\begin{pmatrix} 3 & -10 & 2 \\ -3 & 9 & 0 \\ 0 & -2 & 1 \end{pmatrix}$

(l) $\frac{1}{53}\begin{pmatrix} 5 & -8 & 33 \\ -7 & -10 & 28 \\ -1 & -9 & 4 \end{pmatrix}$

(m) $\frac{1}{2}\begin{pmatrix} 2 & 6 & -5 \\ -2 & -12 & 9 \\ 0 & -2 & 1 \end{pmatrix}$

(n) $\frac{1}{54}\begin{pmatrix} -19 & 4 & 7 \\ 24 & 12 & -6 \\ 4 & 2 & -10 \end{pmatrix}$

(o) $-\frac{1}{20}\begin{pmatrix} -5 & -7 & -1 \\ 5 & 11 & -7 \\ 5 & -1 & -3 \end{pmatrix}$

(p) $-\frac{1}{57}\begin{pmatrix} 2 & -13 & 9 \\ -33 & 15 & -6 \\ 5 & -4 & -6 \end{pmatrix}$

Exercise 3C

1 (a) $\begin{pmatrix} 1 & 1 \\ 1 & -1 \end{pmatrix}$ (b) $\begin{pmatrix} 1 & 0 \\ 2 & 1 \end{pmatrix}$

(c) $\begin{pmatrix} 3 & 1 \\ -1 & -1 \end{pmatrix}$ (d) $\begin{pmatrix} 1 & 1 \\ 3 & 1 \end{pmatrix}$

2 (a) $\begin{pmatrix} \frac{1}{2} & \frac{1}{2} \\ \frac{1}{2} & -\frac{1}{2} \end{pmatrix}$

(b) $\begin{pmatrix} 1 & 0 \\ -2 & 1 \end{pmatrix}$

(c) $\begin{pmatrix} \frac{1}{2} & \frac{1}{2} \\ -\frac{1}{2} & -\frac{3}{2} \end{pmatrix}$

3 (a) $\begin{pmatrix} 0 & 1 \\ 1 & 0 \end{pmatrix}$ (b) $\begin{pmatrix} 2 & 0 \\ 0 & 1 \end{pmatrix}$

(c) $\begin{pmatrix} 0 & 1 \\ 2 & 0 \end{pmatrix}$ (d) $\begin{pmatrix} 0 & 2 \\ 1 & 0 \end{pmatrix}$

4 (a) $\begin{pmatrix} 0 & 1 \\ 1 & 0 \end{pmatrix}$ (b) $\begin{pmatrix} \frac{1}{2} & 0 \\ 0 & 1 \end{pmatrix}$

(c) $\begin{pmatrix} 0 & 1 \\ \frac{1}{2} & 0 \end{pmatrix}$

5 (a) $\begin{pmatrix} 2 & 0 & 1 \\ 0 & 1 & 0 \\ 0 & -1 & 1 \end{pmatrix}$

(b) $\begin{pmatrix} -1 & 2 & -3 \\ 2 & -1 & 4 \\ 3 & 4 & 1 \end{pmatrix}$

(c) $\begin{pmatrix} 1 & 8 & -5 \\ 2 & -1 & 4 \\ 1 & 5 & -3 \end{pmatrix}$

(d) $\begin{pmatrix} -2 & 5 & -4 \\ 4 & -5 & 6 \\ 6 & 3 & 4 \end{pmatrix}$

6 (a) $\begin{pmatrix} \frac{1}{2} & -\frac{1}{2} & -\frac{1}{2} \\ 0 & 1 & 0 \\ 0 & 1 & 1 \end{pmatrix}$

(b) $\frac{1}{4}\begin{pmatrix} -17 & -14 & 5 \\ 10 & 8 & -2 \\ 11 & 10 & -3 \end{pmatrix}$

(c) $\frac{1}{8}\begin{pmatrix} -17 & -1 & 27 \\ 10 & 2 & -14 \\ 11 & 3 & -17 \end{pmatrix}$

(d) $\frac{1}{8}\begin{pmatrix} -38 & -32 & 10 \\ 20 & 16 & -4 \\ 42 & 36 & -10 \end{pmatrix}$

7 (a) $\begin{pmatrix} 1 & -1 & 3 \\ 2 & 1 & 4 \\ 0 & 1 & 1 \end{pmatrix}$

(b) $\begin{pmatrix} 1 & 3 & -2 \\ -2 & -9 & 5 \\ 1 & 10 & 4 \end{pmatrix}$

(c) $\begin{pmatrix} 6 & 42 & 5 \\ 4 & 37 & 17 \\ -1 & 1 & 9 \end{pmatrix}$

(d) $\begin{pmatrix} 7 & 0 & 13 \\ -20 & -2 & -37 \\ 21 & 13 & 47 \end{pmatrix}$

8 (a) $\frac{1}{5}\begin{pmatrix} -3 & 4 & -7 \\ -2 & 1 & 2 \\ 2 & -1 & 3 \end{pmatrix}$

(b) $\frac{1}{25}\begin{pmatrix} 86 & 32 & 3 \\ -13 & -6 & 1 \\ 11 & 7 & 3 \end{pmatrix}$

(c) $-\frac{1}{125}\begin{pmatrix} 316 & -373 & 529 \\ -53 & 59 & -82 \\ 41 & -48 & 54 \end{pmatrix}$

(d) $-\frac{1}{125}\begin{pmatrix} 387 & 169 & 26 \\ 163 & 56 & -1 \\ -218 & -91 & -14 \end{pmatrix}$

9 (a) $\begin{pmatrix} 1 & 4 & -2 \\ 0 & 1 & 3 \\ 2 & 7 & 6 \end{pmatrix}$

(b) $\begin{pmatrix} -2 & 4 & 1 \\ 1 & -9 & 1 \\ 3 & 5 & -2 \end{pmatrix}$

(c) $\begin{pmatrix} 0 & 3 & 22 \\ 3 & 2 & -23 \\ -1 & 3 & -3 \end{pmatrix}$

(d) $\begin{pmatrix} -4 & -42 & 9 \\ 10 & 6 & -5 \\ 21 & -25 & -3 \end{pmatrix}$

10 (a) $\begin{pmatrix} 2 & 1 & 6 \\ -1 & 4 & 6 \\ 3 & -7 & -3 \end{pmatrix}$

(b) $\begin{pmatrix} 1 & 3 & -1 \\ 2 & 1 & 1 \\ 4 & -2 & 3 \end{pmatrix}$

(c) $\begin{pmatrix} -4 & 20 & 27 \\ 6 & -1 & 15 \\ 19 & -25 & 3 \end{pmatrix}$

(d) $\begin{pmatrix} 28 & -5 & 17 \\ 31 & -11 & 23 \\ -23 & 8 & -19 \end{pmatrix}$

Example 3D

1 (a) $-2, 7; \begin{pmatrix} 1 \\ -1 \end{pmatrix}, \begin{pmatrix} 4 \\ 5 \end{pmatrix}$

(b) $-1, 6; \begin{pmatrix} 1 \\ 1 \end{pmatrix}, \begin{pmatrix} 2 \\ -5 \end{pmatrix}$

(c) $0, 6; \begin{pmatrix} 1 \\ 2 \end{pmatrix}, \begin{pmatrix} 1 \\ -4 \end{pmatrix}$

(d) $2, 2; \begin{pmatrix} 1 \\ -1 \end{pmatrix}$

(e) $1, 5; \begin{pmatrix} 1 \\ -1 \end{pmatrix}, \begin{pmatrix} 1 \\ 1 \end{pmatrix}$

(f) $2, 3; \begin{pmatrix} -\sqrt{2} \\ 1 \end{pmatrix}, \begin{pmatrix} 1 \\ -\sqrt{2} \end{pmatrix}$

(g) $1, 2; \begin{pmatrix} 1 \\ 0 \end{pmatrix}, \begin{pmatrix} 1 \\ 1 \end{pmatrix}$

2 (a) $0, -2, -3; \begin{pmatrix} 0 \\ 1 \\ -1 \end{pmatrix}, \begin{pmatrix} -2 \\ 1 \\ 0 \end{pmatrix}, \begin{pmatrix} 1 \\ 0 \\ -1 \end{pmatrix}$

(b) $1, 2; \begin{pmatrix} 1 \\ 1 \\ -1 \end{pmatrix}, \begin{pmatrix} 2 \\ 1 \\ 0 \end{pmatrix}$

(c) $5, 1, 1; \begin{pmatrix} 1 \\ 1 \\ 1 \end{pmatrix}, \begin{pmatrix} 1 \\ 0 \\ -1 \end{pmatrix}, \begin{pmatrix} 1 \\ -1 \\ 1 \end{pmatrix}$

(d) $0, 1, 2; \begin{pmatrix} 2 \\ -1 \\ 1 \end{pmatrix}, \begin{pmatrix} 1 \\ 0 \\ 0 \end{pmatrix}, \begin{pmatrix} 2 \\ 1 \\ 1 \end{pmatrix}$

(e) $-2, 4, 7; \begin{pmatrix} 3 \\ -2 \\ 0 \end{pmatrix}, \begin{pmatrix} 1 \\ 0 \\ 0 \end{pmatrix}, \begin{pmatrix} 24 \\ 8 \\ 9 \end{pmatrix}$

(f) $-1, 1, 1; \begin{pmatrix} 0 \\ 0 \\ 1 \end{pmatrix}, \begin{pmatrix} 2 \\ -2 \\ 1 \end{pmatrix}$

(g) $1, 2, 3; \begin{pmatrix} 1 \\ -1 \\ 0 \end{pmatrix}, \begin{pmatrix} 2 \\ -1 \\ -2 \end{pmatrix}, \begin{pmatrix} 1 \\ -1 \\ -2 \end{pmatrix}$

(h) $-1, 1, 2; \begin{pmatrix} 1 \\ 2 \\ -7 \end{pmatrix}, \begin{pmatrix} 1 \\ 0 \\ -1 \end{pmatrix}, \begin{pmatrix} 1 \\ -1 \\ -1 \end{pmatrix}$

(i) $1, 2, 3; \begin{pmatrix} 0 \\ -2 \\ 1 \end{pmatrix}, \begin{pmatrix} 1 \\ 1 \\ 0 \end{pmatrix}, \begin{pmatrix} 2 \\ 2 \\ 1 \end{pmatrix}$

(j) $-1, 1, 3; \begin{pmatrix} -3 \\ 1 \\ 2 \end{pmatrix}, \begin{pmatrix} 1 \\ 1 \\ 0 \end{pmatrix}, \begin{pmatrix} -3 \\ 1 \\ -2 \end{pmatrix}$

3 (a) $\begin{pmatrix} \frac{1}{\sqrt{2}} & \frac{1}{\sqrt{2}} \\ -\frac{1}{\sqrt{2}} & \frac{1}{\sqrt{2}} \end{pmatrix}$

(b) $\begin{pmatrix} \frac{2}{\sqrt{13}} & -\frac{3}{\sqrt{13}} \\ \frac{3}{\sqrt{13}} & \frac{2}{\sqrt{13}} \end{pmatrix}$

(c) $\begin{pmatrix} \frac{\sqrt{3}}{2} & \frac{1}{2} \\ \frac{1}{2} & -\frac{\sqrt{3}}{2} \end{pmatrix}$

(d) $\begin{pmatrix} \frac{2}{\sqrt{5}} & \frac{1}{\sqrt{5}} \\ \frac{1}{\sqrt{5}} & -\frac{2}{\sqrt{5}} \end{pmatrix}$

(e) $\begin{pmatrix} \frac{2}{3} & -\frac{1}{3} & \frac{2}{3} \\ \frac{2}{3} & \frac{2}{3} & -\frac{1}{3} \\ -\frac{1}{3} & \frac{2}{3} & \frac{2}{3} \end{pmatrix}$

(f) $\begin{pmatrix} \frac{1}{\sqrt{2}} & 0 & \frac{1}{\sqrt{2}} \\ 0 & 1 & 0 \\ -\frac{1}{\sqrt{2}} & 0 & \frac{1}{\sqrt{2}} \end{pmatrix}$

(g) $\begin{pmatrix} \frac{1}{\sqrt{3}} & 0 & \frac{2}{\sqrt{6}} \\ \frac{1}{\sqrt{3}} & \frac{1}{\sqrt{2}} & -\frac{1}{\sqrt{6}} \\ \frac{1}{\sqrt{3}} & -\frac{1}{\sqrt{2}} & -\frac{1}{\sqrt{6}} \end{pmatrix}$

(h) $\begin{pmatrix} \frac{1}{3} & \frac{2}{3} & \frac{2}{3} \\ \frac{2}{3} & -\frac{2}{3} & \frac{1}{3} \\ \frac{2}{3} & \frac{1}{3} & -\frac{2}{3} \end{pmatrix}$

(i) $\begin{pmatrix} \frac{2}{3} & \frac{1}{\sqrt{2}} & \frac{1}{3\sqrt{2}} \\ \frac{1}{3} & 0 & -\frac{4}{3\sqrt{2}} \\ \frac{2}{3} & -\frac{1}{\sqrt{2}} & \frac{1}{3\sqrt{2}} \end{pmatrix}$

(j) $\begin{pmatrix} 0 & \frac{5}{\sqrt{45}} & -\frac{2}{3} \\ \frac{2}{\sqrt{5}} & \frac{2}{\sqrt{45}} & \frac{1}{3} \\ \frac{1}{\sqrt{5}} & -\frac{4}{\sqrt{45}} & -\frac{2}{3} \end{pmatrix}$

(k) $\begin{pmatrix} \frac{1}{2} & \frac{1}{\sqrt{2}} & \frac{1}{2} \\ \frac{\sqrt{2}}{2} & 0 & -\frac{\sqrt{2}}{2} \\ \frac{1}{2} & -\frac{1}{\sqrt{2}} & \frac{1}{2} \end{pmatrix}$

(l) $\begin{pmatrix} 0 & \frac{1}{\sqrt{3}} & -\frac{2}{\sqrt{6}} \\ \frac{1}{\sqrt{2}} & \frac{1}{\sqrt{3}} & \frac{1}{\sqrt{6}} \\ -\frac{1}{\sqrt{2}} & \frac{1}{\sqrt{3}} & \frac{1}{\sqrt{6}} \end{pmatrix}$

(m) $\begin{pmatrix} \frac{1}{3} & \frac{2}{3} & \frac{2}{3} \\ -\frac{2}{3} & \frac{2}{3} & -\frac{1}{3} \\ -\frac{2}{3} & -\frac{1}{3} & \frac{2}{3} \end{pmatrix}$

(n) $\begin{pmatrix} \frac{1}{\sqrt{14}} & \frac{1}{\sqrt{3}} & \frac{5}{\sqrt{42}} \\ \frac{2}{\sqrt{14}} & \frac{1}{\sqrt{3}} & -\frac{4}{\sqrt{42}} \\ -\frac{3}{\sqrt{14}} & \frac{1}{\sqrt{3}} & -\frac{1}{\sqrt{42}} \end{pmatrix}$

Exercise 4A

1 (a) $3\mathbf{k}$ (b) $-2\mathbf{j} + 2\mathbf{k}$

   (c) $2\mathbf{k}$ (d) $\mathbf{i} - 3\mathbf{j} - 5\mathbf{k}$

   (e) $10\mathbf{i} + 7\mathbf{j} + \mathbf{k}$ (f) $\mathbf{0}$

   (g) $4\mathbf{i} - 2\mathbf{j} + 3\mathbf{k}$ (h) $2\mathbf{i} - 5\mathbf{j} - 4\mathbf{k}$

   (i) $8\mathbf{i} - 8\mathbf{j} - 8\mathbf{k}$ (j) $-8\mathbf{i} - 19\mathbf{j} - 10\mathbf{k}$

2 $\frac{1}{\sqrt{2}}(\mathbf{i} - \mathbf{j})$

3 $\frac{1}{7}(3\mathbf{i} - 2\mathbf{j} + 6\mathbf{k})$

4 $\frac{7}{\sqrt{3}}(\mathbf{i} + \mathbf{j} + \mathbf{k})$

5 $2\sqrt{2}$

6 (a) -14

   (b) $-8(\mathbf{i} + 3\mathbf{j} + \mathbf{k})$

   (c) $-\frac{1}{\sqrt{11}}(\mathbf{i} + 3\mathbf{j} + \mathbf{k})$

7 (a) $\sqrt{\frac{14}{15}}$ (b) $\sqrt{\frac{23}{55}}$ (c) $\frac{5\sqrt{3}}{9}$

   (d) $\frac{3\sqrt{3}}{14}$ (e) $\frac{\sqrt{7}}{14}$

8 $p = 2, \quad q = -2, \quad \lambda = 2$

9 $a = -1, \quad b = -1, \quad c = -1; \quad -\frac{2\sqrt{2}}{3}$

Exercise 4B

1 $\frac{5\sqrt{2}}{2}$

2 $\frac{13\sqrt{3}}{2}$

3 $\frac{\sqrt{11}}{2}$

4 $\sqrt{11}$

5 $\frac{5\sqrt{10}}{2}$

6 $2\sqrt{14}$

7 $6\sqrt{34}$

8 $\frac{\sqrt{21}}{2}$

9 39

10 (a) $5\mathbf{i} - \mathbf{j} - 7\mathbf{k}, \ 2\mathbf{i} - 8\mathbf{j} + \mathbf{k}$

    (b) (i) $\frac{5\sqrt{3}}{2}$ (ii) $\frac{19}{6}$

11 $\frac{a^2\sqrt{3}}{2}$

12 (a) 24 (b) $6\mathbf{i} + 18\mathbf{j} + 6\mathbf{k}$

    (c) $39.7°$ (d) $3\sqrt{11}$

13 (b) $\frac{1}{2}\sqrt{171}$ (c) $\frac{3}{2}\sqrt{133}$

14 (b) $\frac{8}{3}$

15 12

16 (a) $\frac{3}{2}\sqrt{3}$

    (b) $\pm\frac{1}{\sqrt{27}}(\mathbf{i} - 5\mathbf{j} - \mathbf{k})$

    (c) $\frac{4}{3}$

17 $\frac{1}{3}$

18 (a) $\sqrt{2}$ (b) $\frac{2}{3}$

19 (a) $-2\mathbf{i} - \mathbf{j} + 3\mathbf{k}, \ \mathbf{i} - 3\mathbf{j} + 2\mathbf{k}; \ 7\mathbf{i} + 7\mathbf{j} + 7\mathbf{k}$

    (b) $\frac{7}{2}\sqrt{3}$

    (c) $\frac{7}{3}$

20 (a) $\mathbf{i} + 2\mathbf{j}$

    (b) $2\sqrt{5}, \ \frac{1}{2}\sqrt{5}; \ \frac{5}{3}$

Exercise 4C

1 (a) $\mathbf{r} \times (-\mathbf{i}+\mathbf{j}-2\mathbf{k}) = \mathbf{i}+3\mathbf{j}+\mathbf{k}$

(b) $\mathbf{r} \times (2\mathbf{i}+3\mathbf{j}) = 6\mathbf{i}-4\mathbf{j}+3\mathbf{k}$

(c) $\mathbf{r} \times (2\mathbf{i}+\mathbf{j}-3\mathbf{k}) = 13\mathbf{i}+7\mathbf{j}+11\mathbf{k}$

(d) $\mathbf{r} \times (3\mathbf{i}-2\mathbf{j}+\mathbf{k}) = 7\mathbf{i}+11\mathbf{j}+\mathbf{k}$

2 (a) $\left[\mathbf{r} - \begin{pmatrix} 3 \\ 2 \\ 5 \end{pmatrix}\right] \times \begin{pmatrix} 6 \\ 3 \\ 12 \end{pmatrix} = \mathbf{0}$

(b) $\left[\mathbf{r} - \begin{pmatrix} 1 \\ -2 \\ -1 \end{pmatrix}\right] \times \begin{pmatrix} 1 \\ -2 \\ -3 \end{pmatrix} = \mathbf{0}$

(c) $\left[\mathbf{r} - \begin{pmatrix} -4 \\ -3 \\ 11 \end{pmatrix}\right] \times \begin{pmatrix} 4 \\ 0 \\ -10 \end{pmatrix} = \mathbf{0}$

(d) $\left[\mathbf{r} - \begin{pmatrix} 3 \\ -2 \\ 3 \end{pmatrix}\right] \times \begin{pmatrix} -4 \\ 2 \\ -2 \end{pmatrix} = \mathbf{0}$

3 (a) $\left[\mathbf{r} - \begin{pmatrix} 1 \\ -1 \\ 2 \end{pmatrix}\right] \times \begin{pmatrix} 2 \\ -1 \\ 1 \end{pmatrix} = \mathbf{0}$

(b) $\left[\mathbf{r} - \begin{pmatrix} -1 \\ 3 \\ 1 \end{pmatrix}\right] \times \begin{pmatrix} -1 \\ -3 \\ -4 \end{pmatrix} = \mathbf{0}$

(c) $\left[\mathbf{r} - \begin{pmatrix} 1 \\ 0 \\ -4 \end{pmatrix}\right] \times \begin{pmatrix} 2 \\ 1 \\ -2 \end{pmatrix} = \mathbf{0}$

(d) $\left[\mathbf{r} - \begin{pmatrix} 2 \\ -1 \\ 3 \end{pmatrix}\right] \times \begin{pmatrix} -3 \\ 2 \\ 0 \end{pmatrix} = \mathbf{0}$

4 (a) $\mathbf{r} . (2\mathbf{i}+4\mathbf{j}-\mathbf{k}) = -4$

(b) $\mathbf{r} . (-\mathbf{i}+3\mathbf{j}-4\mathbf{k}) = -22$

(c) $\mathbf{r} . (2\mathbf{i}+3\mathbf{j}-4\mathbf{k}) = 11$

(d) $\mathbf{r} . (-3\mathbf{i}+\mathbf{k}) = -4$

(e) $\mathbf{r} . (6\mathbf{i}+4\mathbf{j}-2\mathbf{k}) = 14$

5 (a) $2x+4y-z = -4$

(b) $-x+3y-4z = -22$

(c) $2x+3y-4z = 11$

(d) $-3x+z = -4$

(e) $3x+2y-z = 7$

7 (a) (i) $\mathbf{r} = \begin{pmatrix} 1 \\ -1 \\ 1 \end{pmatrix} + \lambda \begin{pmatrix} 1 \\ -3 \\ 2 \end{pmatrix} + \mu \begin{pmatrix} -1 \\ 2 \\ -4 \end{pmatrix}$

(ii) $8x+2y-z = 5$

(b) (i) $\mathbf{r} = \begin{pmatrix} 4 \\ 7 \\ -1 \end{pmatrix} + \lambda \begin{pmatrix} 1 \\ 2 \\ 1 \end{pmatrix} + \mu \begin{pmatrix} -2 \\ -9 \\ 4 \end{pmatrix}$

(ii) $17x-6y-5z = 31$

(c) (i) $\mathbf{r} = \begin{pmatrix} 8 \\ 1 \\ -1 \end{pmatrix} + \lambda \begin{pmatrix} -6 \\ 5 \\ -1 \end{pmatrix} + \mu \begin{pmatrix} -5 \\ -4 \\ 1 \end{pmatrix}$

(ii) $x+11y+49z = -30$

(d) (i) $\mathbf{r} = \begin{pmatrix} 2 \\ 0 \\ -3 \end{pmatrix} + \lambda \begin{pmatrix} -1 \\ 4 \\ 2 \end{pmatrix} + \mu \begin{pmatrix} 0 \\ -1 \\ 3 \end{pmatrix}$

(ii) $14x+3y+z = 25$

8 (a) $x+y+z = 3$

(b) $7x-6y-4z = 12$

(c) $3x-4y-6z+13 = 0$

(d) $3x+4y-24z = 12$

9 (a) $(0, -2, 3)$ (b) $(3, 1, 2)$

10 (a) $\mathbf{r} . (4\mathbf{i}-3\mathbf{j}-\mathbf{k}) = 0$

(b) $\mathbf{r} . \begin{pmatrix} 4 \\ -16 \\ 5 \end{pmatrix} = -10$

(c) $\mathbf{r} . \begin{pmatrix} 9 \\ 3 \\ 5 \end{pmatrix} = -14$

11 $x-2y+z = 0$

12 (a) $4x-3y+2x = 6$

(d) $\left(2\frac{1}{2}, 2, 1\right)$

Exercise 4D

1 (a) $\frac{5}{14}\sqrt{14}$ (b) $\frac{12}{23}\sqrt{46}$ (c) $\frac{17}{141}\sqrt{141}$

(d) $\frac{62}{13}$ (e) $\frac{5}{2}\sqrt{6}$

2 (a) $\frac{5}{2}\sqrt{6}$ (b) $\frac{5}{2}\sqrt{14}$ (c) 3

(d) $3\frac{1}{3}$ (e) $\frac{9}{22}\sqrt{66}$

3 $2\mathbf{i}+\mathbf{j}+3\mathbf{k}$

4 (a) $46.4°$ (b) $40.3°$ (c) $67.1°$

5 (a) $27.3°$ (b) $53.5°$ (c) $81.8°$

6 2

7 $\dfrac{18\sqrt{29}}{203}$

8 (a) $\mathbf{r} = \frac{121}{14}\mathbf{i} - \frac{25}{7}\mathbf{k} + \lambda\left(\frac{10}{7}\mathbf{i}+\mathbf{j}+\frac{9}{7}\mathbf{k}\right)$

(b) $\mathbf{r} = -\frac{11}{4}\mathbf{j} - \frac{7}{4}\mathbf{k} + \lambda\left(\mathbf{i}-\frac{1}{2}\mathbf{j}+2\mathbf{k}\right)$

(c) $\mathbf{r} = 20\mathbf{i} - 25\mathbf{k} + \lambda(5\mathbf{i}+\mathbf{j}-7\mathbf{k})$

9 $78.5°$

10 2

11 (a) $x - 2y + z = 0$ (b) $84.5°$

12 $10.6°$

14 (a) $\mathbf{i} + 3\mathbf{j} + 2\mathbf{k}$

(c) $\mathbf{r}.\begin{pmatrix} -1 \\ 3 \\ 7 \end{pmatrix} = 22$

15 (a) $\mathbf{r} = \begin{pmatrix} -4 \\ 4 \\ -1 \end{pmatrix} + \mu\begin{pmatrix} 3 \\ -2 \\ 4 \end{pmatrix}$

(b) $\mathbf{r} = s\begin{pmatrix} 5 \\ -1 \\ -3 \end{pmatrix} + t\begin{pmatrix} -4 \\ 4 \\ -1 \end{pmatrix}$

(c) $\sqrt{\frac{21}{55}}$; $\mathbf{r}.\begin{pmatrix} 3 \\ -2 \\ 4 \end{pmatrix} = 5$

(d) $3x - 2y + 4z = 5$

(e) $\frac{x+4}{3} = \frac{y-4}{-2} = \frac{z+1}{4}$

16 $4x - 2y - 3z = 5$

17 $\frac{1}{5\sqrt{2}}(3\mathbf{i} + 5\mathbf{j} + 4\mathbf{k})$; $3x + 5y + 4z = 30$; $3\sqrt{2}$

18 $\frac{1}{2}\sqrt{2}$; $x + z = 1$

19 (c) $\frac{x}{1} = \frac{y+1}{1} = \frac{z-1}{-5}$

20 $\mathbf{r} = \begin{pmatrix} 2 \\ -1 \\ 2 \end{pmatrix} + s\begin{pmatrix} -3 \\ 1 \\ -1 \end{pmatrix} + t\begin{pmatrix} 4 \\ -1 \\ 2 \end{pmatrix}$

$\mathbf{r} = \begin{pmatrix} 6 \\ 7 \\ -2 \end{pmatrix} + p\begin{pmatrix} -3 \\ 1 \\ -1 \end{pmatrix} + q\begin{pmatrix} 4 \\ -1 \\ 2 \end{pmatrix}$

21 (b) $7\mathbf{i} + 2\mathbf{j} - 6\mathbf{k}$

(c) $\frac{14}{15}$

(d) $\mathbf{r} = \begin{pmatrix} 7 \\ 2 \\ -6 \end{pmatrix} + s\begin{pmatrix} 2 \\ 1 \\ -2 \end{pmatrix} + t\begin{pmatrix} -3 \\ 0 \\ 4 \end{pmatrix}$

22 (a) $4\mathbf{i} + 8\mathbf{j} + 10\mathbf{k}$; $\mathbf{r}.\begin{pmatrix} 4 \\ 8 \\ 10 \end{pmatrix} = 0$; $3\sqrt{5}$

(b) $\mathbf{r} = \begin{pmatrix} -2 \\ 1 \\ 0 \end{pmatrix} + s\begin{pmatrix} 4 \\ 8 \\ 10 \end{pmatrix} + t\begin{pmatrix} 1 \\ 2 \\ -2 \end{pmatrix}$

(c) $\frac{x-1}{-1} = \frac{y-2}{3} = \frac{z+2}{-2}$ $(= \lambda)$

23 (c) $\mathbf{r} = \begin{pmatrix} 2 \\ -2 \\ 3 \end{pmatrix} + \lambda\begin{pmatrix} 2 \\ -1 \\ 1 \end{pmatrix}$

(d) $(-1, -\frac{1}{2}, 1\frac{1}{2})$ (e) $\frac{3}{2}\sqrt{6}$

(f) $\mathbf{r}.\begin{pmatrix} 2 \\ -1 \\ 1 \end{pmatrix} = 9$

24 $\mathbf{r} = \begin{pmatrix} 5 \\ 7 \\ 9 \end{pmatrix} + s\begin{pmatrix} 1 \\ 2 \\ 3 \end{pmatrix} + t\begin{pmatrix} 4 \\ 5 \\ 6 \end{pmatrix}$; $(5, 7, 9)$

25 $\frac{7}{3}$ **27** $3\sqrt{2}$ **28** $3\sqrt{2}$ **29** $\frac{7\sqrt{6}}{9}$

Exercise 5A
1 1.2033 **2** 2.1468 **3** 2.3666 **4** 0.1117
5 5.7600 **6** 1.7545 **7** 0.3365 **8** 2.0483
9 1.0408 **10** 1.9905 **11** 0.4625 **12** 1.7912
13 0.5587
14 (a) 0.4609 (b) 0.4961
15 (a) 1.2213 (b) 1.3508
16 2.5171
17 (a) 2.222 (b) 2.368
18 (a) 0.0525 (b) 0.1103
19 (a) 4.32 (b) 14.9384
20 2.2214

Review exercise
1 $\frac{1}{2} \pm i\frac{\sqrt{3}}{2}$, -1 **3** 9
4 (a) $\lambda\mathbf{i} - 2\mathbf{j} + \lambda\mathbf{k}$ (b) ±2
5 $1 + x - 4x^2 - \frac{13}{3}x^3$
6 $3x^2 + 3y^2 + 10x + 3 = 0$
7 (a) $27°$ (b) $u = -3$, $v = 5$
8 $-3x^2 - 2x^3$
9 8
10 $\sqrt{2}$, $\frac{\pi}{4}$; $2^{\frac{1}{8}}e^{\frac{\pi i}{16}}$, $2^{\frac{1}{8}}e^{\frac{9\pi i}{16}}$, $2^{\frac{1}{8}}e^{-\frac{7\pi i}{16}}$, $2^{\frac{1}{8}}e^{-\frac{15\pi i}{16}}$
11 $2, 5, -2$; $\begin{pmatrix} 2 & 0 & 0 \\ 0 & 5 & 0 \\ 0 & 0 & -2 \end{pmatrix}$
13 (b) $1 - x + \frac{x^2}{2} - \frac{x^3}{3}$
14 $-1, 11, 3$ **15** $3, -2$
16 3, $\frac{1}{7}\begin{pmatrix} 6 \\ 2 \\ -3 \end{pmatrix}$
17 $1 + \frac{3}{2}x + \frac{11}{8}x^2$; 1.51% increase
18 (a) $-\mathbf{i} + \mathbf{j} + 4\mathbf{k}$
(c) $27°$

21 (a) $(1, -2, 3)$

(b) $2x + y - 3z = -9$

(c) $70.9°$

22 (b) $(\sqrt{5}, \sqrt{5} - 1)$

23 (a) $75°$

(b) $\mathbf{r} = \mathbf{i} - 2\mathbf{j} + 2\mathbf{k} + t(\mathbf{i} - 3\mathbf{j} - 3\mathbf{k})$

(c) 2.1

24 (c) $1 + \frac{5}{2}x^2 + \frac{41}{24}x^4$

26 (a) $a = \frac{25}{6}, b = 0, c = -\frac{25}{6}$

(b) $\frac{25}{3}$

27 (a) $\frac{1}{7}\begin{pmatrix} 5 & 2 & -1 \\ -17 & 3 & 2 \\ 2 & -2 & 1 \end{pmatrix}$

(b) $\frac{x}{10} = y = -\frac{z}{3}$

28 $2[x + \frac{x^3}{3} + \frac{x^5}{5}]; |x| < 1; a = \frac{2\sqrt{3}}{3}, b = \sqrt{3}$

30 $\frac{4}{3}, 3$ units^2

32 -2048

33 (a) $1 + \frac{1}{2}x^2 + \frac{3}{8}x^4 + \frac{5}{16}x^6$

(c) 1.41415

34 $\dfrac{(5-\mathrm{i})z - 2\mathrm{i}}{z}$

36 $x + \frac{1}{2}x^2 - \frac{1}{6}x^3$

37 (a) 1

(b) $\frac{1}{\sqrt{21}}\begin{pmatrix} 4 \\ 1 \\ -2 \end{pmatrix}$

(c) $\begin{pmatrix} \frac{1}{\sqrt{14}} & \frac{1}{\sqrt{6}} & \frac{4}{\sqrt{21}} \\ \frac{2}{\sqrt{14}} & -\frac{2}{\sqrt{6}} & \frac{1}{\sqrt{21}} \\ \frac{3}{\sqrt{14}} & \frac{1}{\sqrt{6}} & -\frac{2}{\sqrt{21}} \end{pmatrix}$

(d) $\begin{pmatrix} -1 & 0 & 0 \\ 0 & 8 & 0 \\ 0 & 0 & 1 \end{pmatrix}$

38 (a) $7\mathbf{i} - 7\mathbf{j} - 14\mathbf{k}$

(b) $\mathbf{a}.(\mathbf{b} \times \mathbf{c}) = 0 \quad \therefore A, B, C$ coplanar

39 (b) $2x - \frac{4}{3}x^3 + \dots$

(c) $k = 2$

40 $a = b = 3$

41 $\begin{pmatrix} 3 & 1 & -3 \\ -3 & -2 & 5 \\ 1 & 1 & -2 \end{pmatrix}; \begin{pmatrix} 2 \\ -2 \\ 1 \end{pmatrix}$

42 $2(1+x)\ln(1+x) + (1+x); 2\ln(1+x) + 3;$

$\dfrac{2}{1+x}$

$x + \frac{3}{2}x^2 + \frac{1}{3}x^3 + \dots$

43 $\frac{1}{27}\begin{pmatrix} 6 & -4 & 11 \\ 0 & -9 & 18 \\ 3 & -11 & 10 \end{pmatrix}$

44 $3y' = 8x' - 5$

45 (a) $16\cos^5\theta - 20\cos^3\theta + 5\cos\theta$

(b) $2e^{\frac{\mathrm{i}\pi}{6}}, 2e^{\frac{\mathrm{i}5\pi}{6}}, 2e^{-\frac{\mathrm{i}\pi}{2}}$

47 $4 + \sqrt{2}, 4 - \sqrt{2}$ **48** $y^2 = 4(x+1)$

49 (a) $2x + \frac{3}{2}x^2 + \frac{1}{3}x^3; 0.2155$

(b) 0.463

52 (a) $1 + x - 2x^2 - \frac{7}{6}x^3 + \frac{5}{3}x^4$

(b) 1.08

53 (a) $-2\mathbf{i} - \mathbf{j} - 2\mathbf{k}$ (c) 2

(d) $\frac{8}{3}\mathbf{i} + \frac{4}{3}\mathbf{j} + \frac{8}{3}\mathbf{k}$

54 $x - \frac{1}{2}x^2 + \frac{1}{6}x^3 - \frac{1}{12}x^4; 0.116$

55 (b) $p = \frac{1}{8}, q = \frac{1}{2}, r = \frac{3}{8}$

56 (a) $2(\cos\frac{\pi}{3} + \mathrm{i}\sin\frac{\pi}{3})$

57 $1 + \frac{1}{2}x - \frac{1}{8}x^2 + \frac{1}{16}x^3 - \dots,$

$a = -\frac{1}{2}, b = -\frac{1}{8}$

58 $-\frac{1}{2}x^2 - \frac{1}{12}x^4 + \dots, \frac{1}{2}x^2 + \frac{1}{12}x^4 + \dots$

59 (b) $-10\mathbf{i} + 10\mathbf{j} - 5\mathbf{k}$

(c) $2x - 2y + z = 10$

(d) 15 units^2

61 (a) $-6\mathbf{i} - 4\mathbf{j} + 2\mathbf{k}$

(b) $\mathbf{r}.(-6\mathbf{i} - 4\mathbf{j} + 2\mathbf{k}) = 0$

(c) $(-1, 1, -1)$

63 1.222

64 $1 + \frac{3}{2}x^2 + 2x^3 + \frac{5}{4}x^4; 1.078$

65 (b) $1 - x + \frac{1}{2}x^2$

(c) $\frac{1}{15}$

66 0.035%

67 1.802

69 (a) $\begin{pmatrix} -\frac{1}{2} & -\frac{1}{2} & \frac{5}{2} \\ 0 & -4 & 3 \\ 0 & 0 & 4 \end{pmatrix}$

(b) $\begin{pmatrix} -1 & 1 & -1 \\ 0 & -1 & 2 \\ 1 & -1 & 2 \end{pmatrix}$

(c) $p = 0, q = -11, r = 11$

71 $1 + \frac{1}{2}x^2 + \frac{5}{24}x^4 + \ldots$

72 5   **73** $\frac{22}{3} + \frac{5}{2}\pi$

79 (a)   1.039   (b)   1.055

80 (a) $\sqrt{91}$

      (b) $\mathbf{r}.(9\mathbf{i} + 3\mathbf{j} - \mathbf{k}) = 11$

81 (a)   3.13, 3.27

      (b) $y = 2 + 4x + x^2 - \frac{2}{3}x^3 \ldots$

83 (a) $27°$ (b) $\frac{1}{2}\sqrt{17}$

      (c) $\mathbf{r}.(2\mathbf{i} + 2\mathbf{j} - 3\mathbf{k}) = -6$

      (d) $\frac{5}{\sqrt{17}}$

84 $\frac{1}{2}\begin{pmatrix} 2 & 0 & 0 \\ -x & 1 & 0 \\ x-6 & -1 & 2 \end{pmatrix}$

85 (a) $56 - 33i$, $\frac{16+63i}{25}$

      (b) $2, \frac{\pi}{3}; -512(1 + i\sqrt{3})$

86 (a) $1 - i\sqrt{3}, -2 - 2i\sqrt{3}, \frac{1}{4} + \frac{i\sqrt{3}}{4};$

           $2, -\frac{\pi}{3}; 4, -\frac{2\pi}{3}, \frac{1}{2}, \frac{\pi}{3}$

91 (a) $1 + x + 2x^2 + 2x^3$

      (b)   1.12

92 $\mathbf{r}.(2\mathbf{i} + \mathbf{k}) = 0; \sqrt{5}$ **93** $2 + x - x^2 - \frac{1}{6}x^3$

95 $\mathbf{r}.(2\mathbf{i} + 2\mathbf{j} + \mathbf{k}) = 9$

97 (a) 14.2 units^2

      (b) $\mathbf{r}.(25\mathbf{i} + 9\mathbf{j} - 10\mathbf{k}) = 15$

      (c) $3; \mathbf{i} + \mathbf{j}, \mathbf{i} - 2\mathbf{j} + 2\mathbf{k}, \mathbf{j} + \mathbf{k}$

99 $82.6°$

101 $\mathbf{r}.(\mathbf{i} + \mathbf{k}) = 3; \frac{1}{2}\sqrt{2} \text{ units}^2; \frac{1}{2}\sqrt{2} \text{ units}$

103 (a) $\mathbf{i} - \mathbf{j} - \mathbf{k}$ (b) $-2; \frac{2}{3}\sqrt{3}$

104 (a) $3\mathbf{i} - 6\mathbf{j} + 6\mathbf{k}$ (c)   14

       (d) $\mathbf{r} = 4\mathbf{i} + 3\mathbf{j} + \mathbf{k} + t(\mathbf{j} + \mathbf{k})$

       (e) $4\mathbf{i} + \mathbf{j} - \mathbf{k}$

105 $-9, 18; \frac{1}{3}\begin{pmatrix} 2 \\ 2 \\ -1 \end{pmatrix}, \frac{1}{3}\begin{pmatrix} 2 \\ -1 \\ 2 \end{pmatrix}, \frac{1}{3}\begin{pmatrix} 1 \\ -2 \\ -2 \end{pmatrix}$

106 (a) $-1, 2, 3; \begin{pmatrix} 1 \\ 0 \\ -2 \end{pmatrix}, \begin{pmatrix} 1 \\ -1 \\ 1 \end{pmatrix}, \begin{pmatrix} 1 \\ 0 \\ 2 \end{pmatrix}$

       (b) $x = \frac{y}{2} = \frac{z}{10}$

107 (a) $\cos 0 + i \sin 0, \cos\frac{2\pi}{5} \pm i\sin\frac{2\pi}{5},$

          $\cos\frac{4\pi}{5} \pm i\sin\frac{4\pi}{5}$

      (b) $(z - 1)(z^2 - 2z\cos\frac{2\pi}{5} + 1)(z^2 - 2z\cos\frac{4\pi}{5} + 1)$

      (c) $\cos\frac{2\pi}{5} = \frac{\sqrt{5}-1}{4}$

108 1.203

109 (a)   6, 9   (c) $\begin{pmatrix} 2 \\ 0 \\ 1 \end{pmatrix}, \begin{pmatrix} 1 \\ 0 \\ 2 \end{pmatrix}$

110 (a) $-1, 3$ (b) $\begin{pmatrix} 1 \\ -2 \\ -4 \end{pmatrix}$

      (c) $\begin{pmatrix} 2 & 2 & 1 \\ 3 & -1 & -2 \\ -1 & 1 & -4 \end{pmatrix}$

111 (a) $\frac{1}{7}\begin{pmatrix} -14 & -a & 7 \\ 7 & 0 & 0 \\ 0 & 1 & 0 \end{pmatrix}$

      (b) $3\mathbf{i} + \mathbf{j} - \mathbf{k}$

113 2.042

114 $y = 3 + 2x + 3x^2 - \frac{7}{3}x^3$

115 $2 + 4x + x^2 - \frac{2}{3}x^3$

116 $4\mathbf{i} + 7\mathbf{j} - 5\mathbf{k}$

      (a) $2\mathbf{i} + \mathbf{j} - 5\mathbf{k}$

      (b) $\sqrt{56}$

117 (a) $\begin{pmatrix} 1 & -1 & 2 \\ 0 & 1 & -1 \\ 0 & 0 & 1 \end{pmatrix}$

      (b)   9   (c) $(\frac{1}{3}, -\frac{1}{3}, 1)$

Examination style paper FP3

2 $[-2, 2]$

3 $1 + x + \frac{1}{2}x^2 + \frac{1}{2}x^3 + \frac{3}{8}x^4$

4 $2e^{\frac{\pi i}{12}}, 2e^{\frac{7\pi i}{12}}, 2e^{-\frac{5\pi i}{12}}, 2e^{-\frac{11\pi i}{12}}$

5   0.9801

6   (a) $1, -2, 3$

      (b) $-\mathbf{i} + \mathbf{j} + \mathbf{k}, 11\mathbf{i} + \mathbf{j} - 14\mathbf{k}, \mathbf{i} + \mathbf{j} + \mathbf{k}$

7   (a) $(1, -1, 2)$

      (b) $\frac{1}{2}\sqrt{14}$

      (c) $4\mathbf{i} - \mathbf{j} + 2\mathbf{k}$

List of symbols and notation

The following notation will be used in all Edexcel examinations.

$\in$	is an element of
$\notin$	is not an element of
$\{x_1, x_2, \ldots\}$	the set with elements $x_1, x_2, \ldots$
$\{x : \ldots\}$	the set of all x such that $\ldots$
$\mathrm{n}(A)$	the number of elements in set A
$\varnothing$	the empty set
$\mathscr{E}$	the universal set
A	the complement of the set A
$\mathbb{N}$	the set of natural numbers, $\{1, 2, 3, \ldots\}$
$\mathbb{Z}$	the set of integers, $\{0, \pm 1, \pm 2, \pm 3, \ldots\}$
$\mathbb{Z}^+$	the set of positive integers, $\{1, 2, 3, \ldots\}$
$\mathbb{Z}_n$	the set of integers modulo n, $\{0, 1, 2, \ldots, n-1\}$
$\mathbb{Q}$	the set of rational numbers $\left\{\dfrac{p}{q} : p \in \mathbb{Z}, q \in \mathbb{Z}^+\right\}$
$\mathbb{Q}^+$	the set of positive rational numbers, $\{x \in \mathbb{Q} : x > 0\}$
$\mathbb{Q}_0^+$	the set of positive rational numbers and zero, $\{x \in \mathbb{Q} : x \geqslant 0\}$
$\mathbb{R}$	the set of real numbers
$\mathbb{R}^+$	the set of positive real numbers, $\{x \in \mathbb{R} : x > 0\}$
$\mathbb{R}_0^+$	the set of positive real numbers and zero, $\{x \in \mathbb{R} : x \geqslant 0\}$
$\mathbb{C}$	the set of complex numbers
(x, y)	the ordered pair x, y
$A \times B$	the cartesian product of sets A and B, $A \times B = \{(a, b) : a \in A, b \in B\}$
$\subseteq$	is a subset of
$\subset$	is a proper subset of
$\cup$	union
$\cap$	intersection
$[a, b]$	the closed interval, $\{x \in \mathbb{R} : a \leqslant x \leqslant b\}$
$[a, b)$	the interval $\{x \in \mathbb{R} : a \leqslant x < b\}$
$(a, b]$	the interval $\{x \in \mathbb{R} : a < x \leqslant b\}$
(a, b)	the open interval $\{x \in \mathbb{R} : a < x < b\}$
$y \, R \, x$	y is related to x by the relation R
$y \sim x$	y is equivalent to x, in the context of some equivalence relation
$=$	is equal to
$\neq$	is not equal to
$\equiv$	is identical to *or* is congruent to
$\approx$	is approximately equal to
$\cong$	is isomorphic to

$\propto$	is proportional to
$<$	is less than
$\leqslant$, $\ngtr$	is less than or equal to, is not greater than
$>$	is greater than
$\geqslant$, $\nless$	is greater than or equal to, is not less than
∞	infinity
$p \wedge q$	p and q
$p \vee q$	p or q (or both)
$\sim p$	not p
$p \Rightarrow q$	p implies q (if p then q)
$p \Leftarrow q$	p is implied by q (if q then p)
$p \Leftrightarrow q$	p implies and is implied by q (p is equivalent to q)
$\exists$	there exists
$\forall$	for all
$a + b$	a plus b
$a - b$	a minus b
$a \times b$, ab, $a.b$	a multiplied by b
$a \div b$, $\dfrac{a}{b}$, a/b	a divided by b
$\displaystyle\sum_{i=1}^{n} a_i$	$a_1 + a_2 + \ldots + a_n$
$\displaystyle\prod_{i=1}^{n} a_i$	$a_1 \times a_2 \times \ldots \times a_n$
$\sqrt{a}$	the positive square root of a
$|a|$	the modulus of a
$n!$	n factorial
$\dbinom{n}{r}$	the binomial coefficient $\dfrac{n!}{r!(n-r)!}$ for $n \in \mathbb{Z}^+$ $\dfrac{n(n-1)\ldots(n-r+1)}{r!}$ for $n \in \mathbb{Q}$
$\mathrm{f}(x)$	the value of the function f at x
$\mathrm{f} : A \to B$	f is a function under which each element of set A has an image in set B
$\mathrm{f} : x \mapsto y$	the function f maps the element x to the element y
f^{-1}	the inverse function of the function f
$\mathrm{g} \circ \mathrm{f}$, gf	the composite function of f and g which is defined by $(\mathrm{g} \circ \mathrm{f})(x)$ or $\mathrm{gf}(x) = \mathrm{g}(\mathrm{f}(x))$
$\displaystyle\lim_{x \to a} \mathrm{f}(x)$	the limit of $\mathrm{f}(x)$ as x tends to a
Δx, δx	an increment of x
$\dfrac{\mathrm{d}y}{\mathrm{d}x}$	the derivative of y with respect to x
$\dfrac{\mathrm{d}^n y}{\mathrm{d}x^n}$	the nth derivative of y with respect to x
$\mathrm{f}'(x), \mathrm{f}''(x), \ldots \mathrm{f}^{(n)}(x)$	the first, second, $\ldots$ nth derivatives of $\mathrm{f}(x)$ with respect to x

$\displaystyle\int y \, \mathrm{d}x$	the indefinite integral of y with respect to x
$\displaystyle\int_a^b y \, \mathrm{d}x$	the definite integral of y with respect to x between the limits $x = a$ and $x = b$
$\dfrac{\partial V}{\partial x}$	the partial derivative of V with respect to x
$\dot{x}, \ddot{x}, \ldots$	the first, second, . . . derivatives of x with respect to t
e	base of natural logarithms
e^x, exp x	exponential function of x
$\log_a x$	logarithm to the base a of x
$\ln x$, $\log_e x$	natural logarithm of x
$\lg x$, $\log_{10} x$	logarithm to the base 10 of x
sin, cos, tan cosec, sec, cot	the circular functions
arcsin, arccos, arctan arccosec, arcsec, arccot	the inverse circular functions
sinh, cosh, tanh cosech, sech, coth	the hyperbolic functions
arsinh, arcosh, artanh, arcosech, arsech, arcoth	the inverse hyperbolic functions
i	square root of -1
z	a complex number, $z = x + \mathrm{i}y$
Re z	the real part of z, Re $z = x$
Im z	the imaginary part of z, Im $z = y$
$|z|$	the modulus of z, $|z| = \sqrt{(x^2 + y^2)}$
arg z	the argument of z, arg $z = \arctan\dfrac{y}{x}$
z^*	the complex conjugate of z, $x - \mathrm{i}y$
$\mathbf{M}$	a matrix $\mathbf{M}$
$\mathbf{M}^{-1}$	the inverse of the matrix $\mathbf{M}$
$\mathbf{M}^{\mathrm{T}}$	the transpose of the matrix $\mathbf{M}$
det $\mathbf{M}$, $|\mathbf{M}|$	the determinant of the square matrix $\mathbf{M}$
$\mathbf{a}$	the vector $\mathbf{a}$
$\overrightarrow{AB}$	the vector represented in magnitude and direction by the directed line segment AB
$\hat{\mathbf{a}}$	a unit vector in the direction of $\mathbf{a}$
$\mathbf{i}, \mathbf{j}, \mathbf{k}$	unit vectors in the directions of the cartesian coordinate axes
$|\mathbf{a}|$, a	the magnitude of $\mathbf{a}$
$|\overrightarrow{AB}|$, AB	the magnitude of $\overrightarrow{AB}$
$\mathbf{a} \cdot \mathbf{b}$	the scalar product of $\mathbf{a}$ and $\mathbf{b}$
$\mathbf{a} \times \mathbf{b}$	the vector product of $\mathbf{a}$ and $\mathbf{b}$

Index